Solar Access Law

Protecting Access to Sunlight for Solar Energy Systems

Solar Access Law

Protecting Access to Sunlight for Solar Energy Systems

Gail Boyer Hayes

An Environmental Law Institute Book

prepared for the

Office of Policy Development and Research
U.S. Department of Housing and Urban Development

in cooperation with the

U.S. Department of Energy

Ballinger Publishing Company □ Cambridge, Massachusetts
A Subsidiary of Harper & Row, Publishers, Inc.

International Standard Book Number: 0-88410-091-X

Library of Congress Catalog Card Number: 79-9392

Printed in the United States of America

Library of Congress Cataloging in Publication Data

Hayes, Gail Boyer.
Solar access law.

"An Environmental Law Institute book."
Bibliography: p.
Includes index.
1. Solar access rights–United States. 2. Solar energy–Law and legislation–United States. I. United States. Dept. of Housing and Urban Development. Office of Policy Development and Research. II. United States. Dept. of Energy. III. Title.
KF5698. 7. H38 343'.73'092 79-9392
ISBN 0-88410-091-X

Contents

Figures

Tables

Preface

> If an ice bridge two miles in diameter could be built, spanning the immense distance of 93,000,000 miles from Sun to Earth, and, if by some means, the whole of the Sun's radiation could be concentrated upon this ice bridge, in one second it would be water and in seven seconds more it would be dissipated into vapour. Though the energy received at the Earth's surface amounts to over 4,000,000 horsepower per square mile, to harness it effectively is difficult.
>
> – ENCYCLOPEDIA BRITTANICA, 15th ed. macropaedia, vol. 17, s.v. "Sun," p. 801

The power of the sun is awesome. But it can be blocked by something as seemingly fragile as an aspen leaf. If we are to harness solar energy, barriers between the sun and solar collectors must be prevented. The Department of Housing and Urban Development asked us to determine the best means of protection.

With great enthusiasm and determination we began investigating ways to protect solar access. Our original goal was to draft a single "best" law that would assure ample sunlight for all solar energy systems. This hope of a single law faded rapidly as we tangled with the complexity of the issues and the possible economic and land use consequences of shadow prevention. Solar energy systems can cause a quiet revolution in our urban and suburban landscapes. Protecting solar collectors' access to direct sunlight will slowly change the skylines and vegetation patterns of existing communities. In new developments, the impact of solar access protection on city planning and on architecture will be even more dramatic. Careful preparation and public education are essential to the success of this revolution.

Approach after approach to solar access protection was examined and discarded. During this process, the help of consultants and reviewers was extremely valuable. Many dedicated individuals spent hours scrutinizing our efforts. Hundreds of changes were suggested and incorporated.

As the project progressed, we learned that it was premature to ask which solar access law is best. We asked instead which types of laws are most promising, which are inferior, and what factors a community should consider when shopping for solar access protection.

Gail Boyer Hayes
Project Director
Environmental Law Institute

Acknowledgments

Many individuals and organizations from all parts of the country made important contributions to this report. The project director particularly wishes to thank those who contributed original materials used in the writing of this report: William A. Thomas, research attorney for the American Bar Foundation; John J. Costonis, professor of law at New York University School of Law; Team Four Inc. Urban Design and Planning as represented by S. Jerome Pratter, principal legal counsel, Daniel R. Mandelker, special consultant to Team Four Inc. and Howard A. Stamper Professor of Law at Washington University School of Law, and Gary H. Feder, legal associate. Current and former staff attorneys at the Environmental Law Institute who helped with the writing or research include Robert A. Friedrich, Richard A. Liroff, Helen Cohn Needham, Joe W. Russell, Jr., and Grant P. Thompson. Randall W. Scott reviewed the early drafts.

Frank O. Meeker III, an ELI attorney with an engineering background, masterminded and authored the technical sections and contributed greatly to all parts of the report. Special thanks are due John J. Costonis for his contributions to the sections on transferable development rights and the taking issue; to William A. Thomas for his analysis of water law, help in developing criteria for evaluation, and material on the federal role; and to Team Four for their work on zoning law, eminent domain, and comprehensive planning.

A technical advisory panel lent their interdisciplinary expertise and practical experience to the project: Duncan S. Bremer, AIA,

AIA Research Corporation; John J. Costonis; Duncan Early, American Planning Association; Gilbert L. Finnell, Jr., professor of law at the University of Houston Law School; Harvey A. Harris, attorney at law, Stolar, Heitzmann & Eder; Lane H. Kendig, director, Lake County (Illinois) Department of Planning, Zoning and Environmental Quality; Albert Ledgard, partner, Wilkes & Artis; Daniel R. Mandelker; Alan Miller, Natural Resources Defense Council; S. Jerome Pratter; Philip Soper, professor of law, University of Michigan Law School; and William A. Thomas.

A draft report of 500 pages was reviewed by more than 20 additional persons selected for their expertise in solar law and related fields. ELI is deeply grateful for the amount of time these experts spent scrutinizing the drafts: Ralph Becker, Jr., attorney, Salt Lake City; Dr. Roger H. Bezdek, chief, Commercialization Planning and Assessment, Conservation and Solar Applications, U.S. Department of Energy; Margaret Lane, U.S. Department of Energy; Nancy Carlisle, Solar Planning Office–West; Jan Chatten-Brown, assistant city attorney, Los Angeles; John R. Ebert, research associate, New York Assembly Research Staff–Minority; Dr. Melvin Eisenstadt, president, Mel Eisenstadt and Associates, Inc.; Clyde O, Fisher, Jr., counsel, Northeast Utilities; Donald G. Hagman, professor of law, University of California at Los Angeles; Linda J. Heller, assistant regional counsel, U.S. Department of Energy, Region II; Seth D. Friedland, legal consultant, U.S. Department of Energy, Region II; Karin H. Hillhouse, private consultant, free-lance writer; Robert K. Huffman, attorney at law, Miller & Chevalier; Sandy F. Kraemer, attorney at law, Kraemer and Kendall; Gerald Mara, demonstration data manager, National Solar Heating and Cooling Information Center; James Matthews, National Conference of State Legislatures; Richard A. Morgan, Oregon Department of Energy; William C. Osborn, Solar Action Office, Commonwealth of Massachusetts; Kenneth A. Rubin, counselor at law, Morgan, Lewis & Bockius; Dennis Schiffel, Solar Energy Research Institute; Robert Odland, senior planner, Solar Energy Research Institute; Richard G. Stein, partner, Richard G. Stein and Partners, Architects; Grant P. Thompson, The Conservation Foundation; Arnold Wallenstein, Northeast Solar Energy Center, Northern Energy Corporation; and Gregory Wheatland, legal counsel, California Energy Resources Conservation and Development Commission.

When so many persons contribute to a study, the problems of coalescence are enormous. Some sections went through more than a dozen drafts; the overall organization was changed several times. Victoria A. Hays came to our rescue in the eleventh hour to track, assemble, and review the many pieces. Without Elizabeth L. Wirtz's excellent secretarial skills the effort would have been doomed from the beginning. Ms. Wirtz also prepared our illustrations. Michael J. Fine, typist extraordinaire as well as urban planner, regularly lent use of his intelligent fingers. The calm professionalism of Linda B. Anderson, office manager, assured us dawn-to-dusk support staff services when needed. Margaret Hough of ELI did a final copy edit; Scott Custin lent his production and design expertise. Nina Graybill, Beth McPherson, and Monique Rothschild of the Franklin Institute's National Solar Heating and Cooling Information Center edited for ideas and structure as well as grammar. And the special genius of Esther M. Tepper, legal assistant and editor, untangled the many reincarnations of footnotes and wrote early drafts of some sections.

Most of all ELI thanks David Engel, an attorney and director of HUD's market development effort in the residential solar demonstration program. As the government technical representative, he kept asking: Is that useful? Will people really read that? Can nonlawyers understand this? Will it actually work? If the answers are yes, then credit must go to all who are mentioned above.

Overview and Conclusions

Scope and Purpose

This study evaluates legal strategies to assure present and future solar energy system owners that they will receive sunlight on their collectors. It focuses on protecting solar access in developed urban and suburban areas. Radically new types of laws, as well as familiar tools such as zoning, are examined as possible protective measures. Recommendations are made on which types of laws we at the Environmental Law Institute think will work best.

It is anticipated that this work will be of most use to legislative staff members, urban planners, and local officials who are trying to develop effective, constitutional, and politically acceptable approaches to ensuring solar access. Attorneys representing clients with solar access problems and persons teaching courses on solar energy law and policy should also find this a helpful resource. Because of the diversity of the intended audience, some sections may seem either too technical or too general for particular readers. The appendices partially alleviate this problem; they contain information that will be useful only to certain readers.

For purposes of this study, it is assumed that solar access protection is needed. Engineers and scientists, not lawyers, must quantify the potential problem. As lawyers we can only note that it seems reasonable to expect disputes to occur as more collectors are installed. A further assumption is that it is desirable to provide reasonable solar access for on-site solar energy systems. While centralized and neighborhood-scale systems may be important in the future, they are not dealt with here. We also presume that people can protect themselves from shadows cast by things on their own property, and that the law should concern itself only with shadows that cross lot lines.

A final premise of the report is that most communities will want to restrict the development of land as little as possible. In other words, they will turn to solar access approaches that provide the most sunlight consistent with placing the least possible restrictions on lots near collectors.

This study does not explain how to protect solar access in new communities where streets, lot sizes, and structure orientation are still undetermined, and where large parcels have not been subdivided and sold to many individuals. Although the access laws described here could apply to new developments as well as old, there are options already at hand that can be used to provide adequate solar access in future developments. For information on solar access in new residential communities, readers are referred to two studies prepared concurrently with this one: the American Planning Association's *Protecting Solar Access for Residential Developments: A Guidebook for Planning Officials* [hereinafter *Guidebook for Planning Officials*], and *Site Planning for Solar Access: A Manual for Developers and Site Planners* [hereinafter *Manual for Developers*].[1]

Conclusions

The most important conclusions of this study are:

- □ There is no single "ideal" solar access law. Different communities or parts of communities may need different legal approaches.
- □ Solar access protection should be part of overall plans to guide land use and to conserve energy.
- □ Because local needs and conditions vary so greatly, it appears that solar access protection can generally be best handled at the local level. The type of solar access law that is best for a given area will depend not only on such physical factors as latitude, climate, architecture, and density, but also on social and economic considerations.

1. (Washington, D.C.: U.S. Government Printing Office, 1979).

- Localities can do much to protect solar access even if no new legislation is enacted at the state level.
- The major legal constraints on solar access law are, in order of importance, the constitutional requirements of public purpose, of equal protection, and of payment of compensation if property is so severely regulated that due process requirements are violated. It is less likely that a solar access law will be found unconstitutional if it is supported by legislative findings of necessity and solar access protection is made part of comprehensive land use plans. Courts' perceptions of the severity of the energy crisis may influence their attitudes on whether public purpose requirements are met.
- Poorly conceived solar access laws could have adverse effects on development patterns, property values, and public attitudes toward solar energy. Well-written access laws can avoid these problems.
- Approaches based on the comparative time of development or use (such as water law analogy) are less fair in their impact, and less certain in their constitutionality, than are controls based on zoning and comprehensive planning.
- Sweeping solar access laws should not be enacted until governments gather empirical data as a basis for their choice of law.

Data Needed

An important part of this study was to determine what governments do not know but must learn before enacting comprehensive solar access laws. Data are needed on:

- The amount of solar access that will be lost each year (in any given community) if no legal protection is implemented. "Lost" does not refer just to the shading of actual collectors, but to the shading of appropriate sites for future collectors.
- The amount of solar access actually available in typical development situations within a city. In other words, how much do structures and vegetation already shade potential collector sites (rooftops, south walls, yards).

- ☐ Available solar energy in cities throughout the United States taking local weather and air pollution into account.
- ☐ The relationship between amounts of protected access and the density of construction. Providing overly generous solar access might encourage urban sprawl, but how much is too much is not known.
- ☐ The impact of different types of vegetation on solar access. For example, cities will need to know the dates local deciduous trees drop their leaves, and the amount of sunlight blocked by their bare branches and twigs.
- ☐ The direct and indirect costs to individuals and to the public of alternative methods of solar access protection.
- ☐ The potential market for retrofitting existing structures with different types of solar energy systems. Market prospects will change along with solar technology and the relative costs of solar and competing energy and associated systems.

Additional general research on solar access laws themselves is not recommended at this time. When communities have access to the data described above, they may then require legal advice on selecting and adapting solar access laws that will meet their particular needs. After local and state solar access laws are enacted, they should be carefully monitored, and their successes and failures communicated to others working in the field.

Spectrum of Options

Communities may choose from a wide range of options to protect solar access. The following list gives a rough idea of possible actions that merit consideration. Some could be undertaken immediately; others will require the collection and evaluation of data. A community may choose one option for one area and a different option for another because of varying conditions (such as use and density). The options are not generally mutually exclusive. The first options require the least government activity; the ones at the end of the list require the most.

Do nothing. The first option is for the local government to do nothing at all. This may be wise, for example, where vegetation is scarce and existing land use controls give adequate protection: An area zoned for large lots with only two or three low-rise dwelling units per acre, for example, may need no further access protection.

Or, where nearly all sites for collectors are already shaded, there may be no access worth protecting. Some climates are so temperate that almost no heating and cooling of buildings is required and the little access required for solar heating of water is already ensured. But in this situation, as in most others, photovoltaics (generating electricity directly from sunlight) or some other technology may eventually be useful. Once access to sunlight is lost, it may be difficult to recover.

Facilitate private agreements. Even without governmental action, private parties can try to buy solar access easements from their neighbors to prevent shading. Many states have considered or enacted legislation that specifically authorizes private solar access easements. Such legislation is advisable because it removes any doubt about whether the local land records office will record solar easements on the titles of the properties involved. Unless the easements are recorded, they would probably not bind subsequent owners of the burdened properties.

Local governments can encourage private arrangements by providing standard forms and guides for property owners and their attorneys to use in preparing the easements. Such aids could greatly reduce the expenses involved. Tax advantages are another possibility; the value of an easement could be excluded from the assessed valuation of the property where the solar collector is located.

Protect solar access only in new residential developments. The American Planning Association's *Guidebook for Planning Officials* and *Manual for Developers* contain suggestions on how existing laws and restrictive covenants can be used to provide solar access in such developments.

Remove barriers in existing land use controls. Regulations on heights, orientation, setbacks, accessory structures, and subdivisions should be examined to find if they unreasonably restrict the addition of solar equipment to existing structures,[2] and solar energy systems should be made an explicitly permitted use in all zones. New laws should prohibit future aesthetic restrictions (both public and private)

2. *See* Mich. H.B. 4493, 4494, and 4495 (1978). The Colonie, N.Y., zoning board decided not to grant a variance to allow a collector in a front yard. An Albany, N.Y., Supreme Court judge upheld this decision. D'Aurio v. Board of Zoning Appeals, 92 Misc. 2d 898 (1978). Determining which restrictions are unreasonable will be difficult—ELI neither condones nor condemns the outcome of the Albany conflict. Readers with similar problems may want to read the imaginative judicial reasoning in this opinion.

from unreasonably restraining solar use. Variance and permit requirements should also be revised, where necessary and possible, to encourage the use of solar equipment.

Revise height definitions. The height regulations in zoning ordinances appear to give more solar access protection than they actually do. If a building has a sloped roof, the legal height is usually half-way up the slope of the roof (a 70-foot A-frame could thus have a legal height of only 35 feet). A locality could improve solar access protection by simply redefining height of a building as the distance from grade to the highest part of the structure. Another problem with typical height definitions is that they allow exceptions that can cast large shadows. Exclusions should be revised, where possible and necessary, to protect access.

Enact state level enabling legislation. A state wishing to encourage solar energy use should pass legislation stating that public policy favors the development of solar energy and that local governments may use their police powers and their power of eminent domain to provide solar access. More specifically, it would be a sensible policy to enable local governments to use their zoning, subdivision, and other land use controls to protect solar access through the regulation of both structures and vegetation.

Control shadows from public vegetation. In areas where shadows from public vegetation are a significant threat to solar energy users, state and local governments may wish to require the planting and pruning of publicly owned vegetation (such as street trees) in a manner that will not destroy solar access.

Add solar access element to comprehensive plan. Protecting solar access will have an impact on urban growth patterns and on many of the goals that are part of comprehensive plans. It will take time and careful study to revise comprehensive plans to integrate solar access with other community goals. This effort will not protect solar access by itself (because comprehensive plans are not self-implementing), but it is a key precursor to most of the options that follow.

Consider solar access in impact statements. The federal government requires environmental impact statements (EISs) to be prepared before many types of public activities can be undertaken. Many states and localities have similar requirements. In some states and localities such statements also have to be prepared when private projects are undertaken. The impact on solar access is something that could be considered during the EIS process.

Use eminent domain to back up private arrangements. In certain situations, such as when urban redevelopments involving many structures are planned, it may be possible to force private arrangements to be made. For instance, if the vast majority of affected landowners in a neighborhood voted to protect solar access, those voting against it might be required to accept compensation in place of development rights.

Control private vegetation. Where structures are so sized and spaced that they do not threaten solar access but where private vegetation may cause shading problems, controls over vegetation may be all that are needed. Controls over private vegetation could be imposed either through the police power or through nuisance law.

Declare shadows on collectors a nuisance. A legislature could declare shadows on collectors to be a public or private nuisance. The state's attorney or the owner of a shaded collector could then sue to have the nuisance shadow abated. If a nuisance approach is used, it is probably best limited to shadows cast by vegetation.

Use transferable development rights (TDR). TDR can sometimes be used when a community does not want to use its eminent domain power to condemn and pay cash for development rights but still wants to compensate the regulated person. Under TDR, airspace development rights are treated as separate from ownership of the land itself. For example, a city could prohibit a property owner from adding four stories to a building that would shade a solar collector, even though the proposed structure would still be within zoning limits. Instead, the restricted property owner would be allowed to sell the right to develop an equivalent amount of area to someone owning property in a "transfer area."

Rezone. In many neighborhoods, structures are not built up to allowed heights or densities but there is no interest in additional construction. In these situations, it may give collector owners peace of mind if a city rezoned the area to reflect actual conditions. A community might also consider extending height limits to apply to vegetation.

Record solar collectors. If a new law so provided, people owning or about to install a collector could apply to the local government for permission to record the existence of their collector in local land records. If recordation were allowed, the local government would issue no permits for the development of nearby land that would result in the shading of the recorded collector. Neighbors could be required to prune trees planted after the collector was recorded so that they would not grow to shade the collector. Solar energy system owners should be required to design and locate collectors so as to reduce the imposition on neighbors.

Adopt solar envelopes. Localities could use their zoning powers to create solar envelopes. Existing zoning could be left undisturbed as the envelopes could be imposed in overlay zones. Most zoning controls set limits on the height, the number of stories, or the floor-to-area ratio of structures on particular properties. It is possible to visualize the result as a box placed roughly over the center of a lot; all building must fit within this box. Solar envelopes are a more elaborate version of this device. They would cover the entire lot and would apply to vegetation as well as to structures. And they might look more like imaginary tents, with sloping roofs, than like rectangular boxes (the sizes and shapes of envelopes can be adjusted to reflect conditions in a particular area).

Enact state-wide solar access laws. Most solar access legislation proposed thus far takes the form of laws that would establish a uniform state-wide approach to solar access. The laws may be mandatory, or may allow localities to opt out. ELI prefers local experimentation and control, however, because local conditions vary so greatly and no proposed statewide bill we have seen adequately takes this into account.

Enact federally mandated solar access laws. Bills have been introduced in the U.S. Senate and House of Representatives that would give all collector owners a right to sunlight. Although such a law may be constitutionally possible under the commerce and defense powers, it would invade an area of law that traditionally belongs to states and localities. The great diversity of conditions in the fifty states argues against a single federal approach. But a national emergency making a quick transition to renewable or on-site energy sources necessary might possibly justify such a federal law.

Enact absolute access laws. To provide the maximum amount of solar access that is physically possible, laws would have to forbid property owners from allowing structures and vegetation to cast shadows beyond the boundaries of their lots (presumably for specified hours of the day). Structures or vegetation that did cast such shadows would be declared nonconforming uses, and would be allowed to remain for only a given number of years. For such laws to be valid, it would be necessary to show they were backed by an extremely convincing public purpose. Such a purpose cannot presently be shown for existing development. In some new low density developments, however, the option of absolute solar access may be feasible, if limited to specified hours of the day.

Steps to Action

Governments concerned about solar access protection might take the following steps:

Determine extent of shading problems. A government should first determine whether the shading of solar collectors is, or is likely to become, a local problem. It may not be a problem where structures and vegetation are scarce compared to land area or where citizens have no interest in using solar energy.

Determine if shadow prevention is wise. If shading is a problem, the government should ask if laws preventing shadows are in the public interest. Although this would generally be the case, there often may be situations within each city where other goals would be inconsistent with, and more important than, solar access or where private arrangements would be more appropriate than public protection. Examples include central city areas with uneven building heights (where extremely dense development is wanted) or hot climates where shading of potential collector sites saves more energy annually than solar collectors would.

Pass enabling legislation, remove barriers, facilitate private agreements, revise comprehensive plans. If solar access legislation appears wise, the minimum action that state or local governments should take is to encourage private arrangements for solar access and remove barriers to solar energy systems in existing land use controls (including building codes). State-level legislation enabling local governments to use their police and eminent domain powers to protect solar access is recommended. Localities should amend ordinances and regulations to require the consideration of impact on solar access as a prerequisite to land development permit decisions (e.g., before a variance is granted). Comprehensive plans should be amended to include consideration of solar access, and possible conflicts between solar access provision and other important public goals should be identified.

Identify sources of shadows. It may be that only trees or only buildings will cast problem shadows. If so, solar access controls may have to deal only with one. This may affect the choice of a legal mechanism for control. A further step would be to find out whether the shadows will come from public land, private land, or both. The government can narrow the controls accordingly.

Identify "typical" developmental situations and determine amount of sunlight available. Topography; height, bulk, and location of structures and vegetation (as well as the species of deciduous vegetation); and the orientation of streets and structures are among factors governments should consider when identifying typical developmental situations. In some areas generalization should be quite easy. (There will be spots, however, that defy categorization.) Governments can then determine how much sunlight is available in these typical situations at various locations on lots: rooftops, south walls, and yards. Aerial photographs may be useful to a community planning for solar access.[3] Photographs of a few residential communities have been studied to determine whether shading problems may exist. It has generally been concluded that in many settings rooftop shading problems are not severe.[4] But photographic evidence must be regarded with some suspicion. First, it is important that series of photographs be taken or that a computer analysis of the photographs be performed that shows the shadows' hourly and daily movement over the course of the year. Second, vegetation growth must be kept in mind; in many areas trees and shrubs will not have attained their full size.

3. Gary Angelici and Nevin A. Bryant studied the northwest part of the Los Angeles metropolitan region to learn how much solar access is available on roofs. They determined land cover types from satellite imagery, and selected a sample site for each land cover type. Then the percentage of each site occupied by flat or south-facing rooftops was found (low altitude photographs were used in this process). "Surprisingly, the results showed that . . . on the average over 120% of existing power demand could be met if only half the potentially usable rooftop area were utilized." SOLAR POTENTIAL INVENTORY MODELLING (Pasadena, Cal.: Jet Propulsion Laboratory, 1978), p. 12.

4. An examination of aerial photographs of typical residential areas in Colorado Springs, Colorado, showed the vast majority of roofs free from shade from 9:00 A.M. to 3:00 P.M. JAMES D. PHILLIPS, ASSESSMENT OF A SINGLE-FAMILY RESIDENCE SOLAR HEATING SYSTEM IN A SUBURBAN DEVELOPMENT SETTING, Solar Heated Residence Annual Research Report, prepared with the support of the National Science Foundation (NSF-RA-N75078, July 1975), p. 92. Scrutiny of aerial photographs of Long Island, N.Y., revealed that over half of all existing buildings checked were suitable for solar water heating and that all new buildings could be equipped with solar water heaters. FRED S. DUBIN, ANALYSIS OF ENERGY USAGE ON LONG ISLAND FROM 1975 TO 1995: THE OPPORTUNITIES TO REDUCE PEAK ELECTRICAL DEMANDS AND ENERGY CONSUMPTION BY ENERGY CONSERVATION, SOLAR ENERGY, AND TOTAL ENERGY SYSTEMS, sponsored by the Suffolk County Department of Environmental Control (Suffolk County, N.Y., 1975).

Decide on amount of access to protect. Governments should then decide how much solar access to protect in each typical developmental situation. For instance, if only rooftop access is presently available in the vast majority of cases in a neighborhood, that access would be a reasonable objective of protection. (If south-wall access is enjoyed by only one in twenty property owners, legal and political problems are likely to arise if a community passes a law purporting to protect south-wall access generally.)

Choose legal mechanism. The next step would be for the government to select the best mechanism for achieving the degree of protection needed under local conditions. Because there are so many variables in both situations and in solutions, it is not possible to offer meaningful hypotheticals to cover all likely situations. (The bulk of this study is aimed at helping governments decide which legal mechanism(s) can best meet their needs.)

Two hypothetical cases will be attempted: Metro City needs a comprehensive plan to protect access. Its citizens are enthusiastic about solar energy, and are worried about preserving access for future collectors as well as for existing ones. In most sections of Metro City, both trees and buildings could shade collectors. Activities on both public and private lands need to be controlled.

A survey of Metro City finds that the amount of sunlight presently available varies greatly among neighborhoods—only rooftop access is available downtown (where buildings tend to have flat tops and be of fairly uniform height) but south-wall access is available in most single family residential neighborhoods (if shadows from trees and shrubs on the same lot as the structure are excluded). Although Metro City can generalize about the type and level of solar access protection that is needed for various neighborhoods, it cannot do so for the city as a whole. Therefore, Metro City needs a flexible approach that is easily adapted to differing needs. Solar overlay zones are an attractive possibility. Metro City has a fairly sophisticated planning staff and the population is growing (there is pressure to develop land more intensely). Metro City may decide that solar envelopes are its best option. Although considerable effort must be put into designing envelopes, they are relatively easy to enforce, can apply to both structures and vegetation, and can effectively protect access for future collectors.

In River Town, on the other hand, relatively few persons are interested in using solar energy. Aside from the public trees along residential streets, there is little tall vegetation. Lot sizes and architectural styles very greatly. Streets are winding. The few persons interested in solar are scattered throughout high-rise buildings and in low density areas. Furthermore, River Town has almost no money to spend on planning.

A sweeping solar access law would be inappropriate here. River Town should instead consider facilitating private solar access easements, redefining "height" in local ordinances, planting and pruning public vegetation in a manner that protects solar access, protecting solar access in new developments, and removing barriers presented by existing land use controls.

Tailor approach to local needs. One of the most serious mistakes that a local or state government can make is to copy another government's solar access law exactly. Suggestions are made throughout this study on the types of modifications localities should consider.

Enact the solar access law.

Monitor changes in solar technology. Finally, because solar energy technology is undergoing rapid change, communities should reexamine their solar access laws periodically to ensure they still provide the type of protection that is needed.

Part I
Introduction

The scales of justice are frequently used to weigh the public interest against private interests. The balance is always a delicate one; indeed, the fulcrum moves over time. In the case of solar access laws, the public need to seek renewable nonpolluting fuels must be weighed against private interests in the traditional rights to develop property. Although laws should always seek to maintain a balance, they must be responsive to social and technological change. Today, one important change is the increasing demand to replace fossil and nuclear fuels with renewable energy resources such as solar energy.

One result of a transition to renewable fuels will be a new realization of the value of sunlight, which solar energy systems require to function. Putting legal and economic values on sunlight may seem novel and rather radical to some people, but a brief look at precedents set by other nations helps dispel this notion. In Japan, for instance, sunshine is treasured for its amenities and access to sunshine is protected by law.[1] Even several hundred years ago, Japanese had to pay compensation if they allowed their trees to grow higher than the ridge of their neighbor's roof, or if they wanted to plant or build near cultivated land.[2] Late in 1976, Japan's Building Standards

1. A dramatic demonstration of the high value of sunlight in Japan occurred when a new four-story apartment building reduced the sunshine at a kindergarden playground next door. The Nagoya District Court ordered a real estate firm to pay $4,500 to the 43 children as compensation. HUD CHALLENGE, January 1977, p. 25.

2. J. WIGMORE, LAW AND JUSTICE IN TOKUGAWA, JAPAN, (ed. Kokusai Bunka Shinkokai (1971), 5: 23, 43.

Law[3] was revised to provide guidelines to help local governments resolve sunshine disputes; the most stringent options available to local governments are quite restrictive in their impact on development.

In England, courts have long recognized a legal device called the ancient lights doctrine, under which limited rights may be acquired to a reasonable amount of light through windows to help make life indoors pleasant. Even this doctrine is a relatively new idea compared to other evidence: an inscription on the Temple of Apollo at Pompeii states that the local government (during the duumvirate of Marcus Helonius and Giaus Ignatius Postumus) had to pay a large sum of money to buy the right to block sunlight reaching the temple's precinct.[4]

In comparison to some other nations, then, the United States has greatly undervalued sunlight. But even here there are laws concerning "light and air" that are intended to make crowded, urban life more bearable.

None of these laws protecting sunlight was written with solar energy systems in mind. Governments placed a high value on this natural resource for other reasons: physical and mental health, aesthetics, an amenable environment, and agriculture. As solar access protection laws are passed in the United States, we may enjoy these indirect benefits as well as the primary one of a new source of energy.

The first chapter of this report details the lack of solar access protection, but a lack can only be defined in terms of a need. Chapter 2, therefore, describes the sunlight needs of solar collectors and some of the technical dimensions of sunshine and shading. If, at times, this technical information seems too detailed, readers can be consoled by the thought that they are in good company because "All the greatest exponents of civilization, from Dante to Goethe, have been obsessed by light."[5] Chapter 3 suggests how to evaluate laws written to meet the sunlight needs of collectors and to fill the legal gaps delineated in Chapter 1.

3. Kenchiku Kijun Ho, Law No. 201 of 1950.

4. JOHN SWARBRICK, DAYLIGHT (London: Batsford, 1953), pp. 8–9.

5. KENNETH CLARK, CIVILIZATION (Rugby, England: Jolly and Barber Ltd., 1970), p. 215.

1
When and Why Solar Access Protection May Be Needed

Although there are now only about 40,000 solar buildings in the United States,[1] some authorities predict that solar technology could meet a quarter of the total energy needs of the United States by the year 2000.[2] Regardless of whether such estimates are optimistic or pessimistic, the increasing development of solar energy will call for new kinds of laws to assure solar energy system owners that their solar collectors[3] will have access to enough sunlight.

Collectors must be exposed to direct sunlight to function well. But the latitude of the contiguous United States causes all direct sunlight to strike the ground at less than a vertical angle. Even at latitudes where the sun is directly overhead once or twice a year at noon, sunlight does not come straight down the rest of the time. Thus,

1. This is the estimate of Donald Beattie, formerly the Acting Assistant Secretary of Energy for Conservation and Solar Applications for the U.S. Department of Energy. Most have only solar-heated hot water. SOLAR ENERGY INTELLIGENCE REPORT 4, no. 11 (13 March 1978): 68. The Department of Energy estimated that the then proposed National Energy Act tax credit would stimulate 2.5 million solar installations by 1985.

2. COUNCIL ON ENVIRONMENTAL QUALITY, SOLAR ENERGY: PROGRESS AND PROMISE (Washington, D.C.: U.S. Government Printing Office, Stock No. 041-011-00036-0, April 1978), p. iv.

3. In this discussion, and in the prototype legislation later in the report, "solar collector" means any device or structure or part of a structure that transforms solar energy into thermal, mechanical, chemical, or electrical energy and that contributes significantly to a structure's energy supply. This broad definition includes collectors that are part of "active" solar systems, as well as "passive" system collectors like south-facing windows.

nearly all land can be shaded, in varying degrees, by vegetation and structures on property owned by others. Moreover, our legal system does not generally recognize any right to access to direct sunlight when that light is blocked by activities on others' properties.

Thousands of solar collectors have been installed without the protection of solar access laws. Legal conflicts have thus far been avoided in all but a few cases because collectors have been very carefully sited, and neighboring trees and structures have not had time to increase in size.[4] Recently in New Jersey, however, a property owner installed collectors on a garage before his neighbor was granted a building permit allowing construction that would shade the collector.[5] Many other such disputes seem likely to occur. Concurrent with the development of solar hardware, therefore, is the need for laws to protect access to sunlight in the most equitable manner possible. Because major adjustments to laws often take a long time to implement, this effort should be started immediately. Hastily written, poorly conceived laws, however, could do more harm than they prevent.

Accommodating a New Public Need

One of the most challenging tasks faced by governments has been to try to maintain a reasonable balance among the rights allotted to each citizen. Adjustments are made necessary by political, economic, social, and technological changes, including the advent of solar technology. Resources that were once free now have price tags. As the population increases, the rights of individuals collide with greater frequency.

Traditionally, our courts and legislatures have placed a high value on allowing the free development of property; a man was said to own his property from the center of the earth straight up to the heavens.[6] This attitude has changed slowly as other important

4. According to SELLING THE SOLAR HOME, U.S. Department of Housing and Urban Development Solar Program Report no. 2 (Washington, D.C., 1978), p. 36, however, planning and zoning officials said that conflicts with existing plans and ordinances could occur if solar use becomes more common, and about a third of those polled mentioned the issue of solar rights.

5. For further information, contact Ray Ebert, the frustrated solar collector owner, in Garwood, New Jersey. As neither his city nor state have solar access laws, Mr. Ebert had no legal grounds on which to contest the shading.

6. *See* WILLIAM L. PROSSER, LAW OF TORTS, 4th ed. (St. Paul: West Publishing Co., 1971), § 13, pp. 69–73. Aircraft above certain altitudes are allowed to penetrate this vertical airspace possession. *See* United States v. Causby, 328 U.S. 256 (1946),

societal needs have been recognized. Zoning laws, building codes, and pollution control laws are examples. Even regulations controlling the exterior appearance of property have been upheld.[7]

Two major types of conflicts over solar access laws can be expected: (1) arguments arising from the economic losses that may result from restrictions on structures and vegetation; and (2) disputes over the aesthetic consequences. (The second issue is beyond the scope of this study, but is presently being researched by ELI for HUD.) Legal approaches are recommended by ELI because they are the most likely to serve the national interest in developing solar energy while avoiding excessive individual losses or gains. Furthermore, none of the recommended solutions need result in dramatic changes in the appearance of neighborhoods.

It is commonly supposed that once collectors are purchased and installed, sunlight offers a free energy source. However, there are other costs involved. To provide one person with access to solar energy, it will usually be necessary to restrict what others may do on nearby land. Additional limitations on how land may be used (such as lower maximum building heights) will sometimes mean a direct dollar loss to the restricted property owner. It is important to note, however, that laws or regulations needed to protect solar access will generally impose relatively minor restrictions on land compared to regulations that already exist. Furthermore, if the price of conventional fuel continues to rise, most landowners may be able to save money, in the long run, from laws ensuring their land some solar access.

Just as restricting the size and placement of structures could result in economic costs for some individuals, so could it impose burdens on society. Limiting urban density can cause urban sprawl; sprawl means more energy must be used to move people and commodities greater distances.[8] Increasingly, low density zoning is being

which declared that this airspace is a public and federal highway. Interferences with solar access, however, occur at lower altitudes than those at issue in *Causby*. While there may be trends away from individual ownership of vertical skyspace, this is not so on an international scale. Eight equitorial nations have recently declared that their boundaries extend 22,500 miles into space. Their goal is to own valuable locations for fixed-position satellites. John J. Fialka, *High Frontier—A Satellite Highway in Space,* Washington Star, 4 April 1978, p. A 1.

7. *See,* e.g., Parkersburg Builders Material Co. v. Barrack, 118 W. Va. 608, 191 S.E. 368, 192 S.E. 291 (1937).

8. *See* Corbin Crews Harwood, Using Land to Save Energy (Cambridge, Mass: Ballinger Publishing Co., 1977), pp. 12–16. The resulting increased travel time is

questioned on grounds that it is exclusionary.[9] Almost any type of solar access protection could encourage urban sprawl if it is too restrictive. Sprawl could result, for example, from laws that flatly ban shadows on any installed collector at any time of the day. But for an existing developed area, sprawl should not be increased by reasonable solar access controls.

Sources of Shadows

The direct solar energy reaching collector surfaces may be diminished by:

- Atmospheric conditions (clouds and air pollution)[10]
- Shadows from structures (including all fixed man-made objects)
- Shadows from vegetation.

This study deals with just the latter two categories.

Because vegetation and structures have very different properties, and because their regulation raises different legal and political issues, they are occasionally given individual treatment in this report. It is nonetheless possible to integrate vegetation and structure controls in a single piece of legislation, and the prototype ordinances in this report do this.

Key differences between vegetation and structures should be kept in mind whenever particular solar access laws are considered:

- Human action and usually a permit of some sort are necessary to bring a structure into existence. Vegetation can propagate itself.

also largely nonproductive to society and increases pollution. Looking far ahead to an era of widespread use of photovoltaics, it may be possible to recharge battery-powered vehicles from home solar systems.

9. *See,* e.g., *Issues in Exclusionary Land Use,* in Management and Control of Growth, ed. Randall W. Scott, David J. Brower, and Dallas D. Miner (Washington, D.C.: Urban Land Institute, 1975), 1: 439–519.

10. The effect that water vapor emissions from a proposed plywood plant would have on solar energy is one issue in a current dispute (contact: Bob Curry, professor of environmental geology, University of Montana, Missoula, Montana). *See also,* U.S. Department of Energy and U.S. Environmental Protection Agency, Solar Energy for Pacific Northwest Residential Heating (n.p., May 1978). The following impacts of air pollution on the collection of solar energy are noted: radiation may be absorbed by pollutants, or it may be scattered, particulate matter may deposit on collector surfaces, and pollutants may act as nuclei for condensation and increase clouds or fog. Ibid., pp. II-25 through II-26.

- Vegetation grows in fairly predictable ways, depending upon species and local conditions. Structures may "grow" as well, but usually only after a permit is issued.
- The value of vegetation is nearly always much less than that of structures.[11]
- The life expectancies of the two vary widely: in some developments the trees have much longer expected life spans than the structures. But in other developments the reverse is true.
- It is generally easier to relocate vegetation, or to offer a replacement to be planted elsewhere, than it is to move a building.

Shadows from Structures

Construction in developed areas usually occurs on a parcel-by-parcel basis: an old building is demolished and replaced by a new, often higher, structure. The replacement of buildings generally occurs very slowly. In both urban and suburban environments, additions to existing structures are common. As the prices of construction and transportation rise, more people may elect to renovate or add to existing buildings close to the central city. It is these changes in the urban environment that will create solar access problems. But it is also these slow changes that could result in improved solar access in tomorrow's urban areas.

Structures are usually extensively regulated. Nearly all state zoning enabling acts delegate traditional police powers to local governments. Ordinances controlling the height and placement of structures are common examples of how the police power is used. But present laws regulating the height and placement of structures do not offer explicit protection of solar access. The setback, yard, and height rules in existing zoning laws give solar collector owners only incidental (and occasional) benefits. In low density zones, for example, there are stringent limits on the size and placement of struc-

11. On the other hand, lay persons usually underestimate the value of large trees. J. J. KIELBASO, ECONOMIC VALUE OF TREES IN THE SOUTH'S URBAN ENVIRONMENT (Athens: University of Georgia Press, 1971), pp. 82–94. The International Society of Arboriculture has developed formulas to determine the monetary value of landscape trees, specimen shrubs, and evergreens. Four factors are considered: (1) kind, (2) size, (3) condition, and (4) location. For example, a shade tree that scored high in all four areas with a trunk diameter of 20 inches would be worth $3,142. INTERNATIONAL SOCIETY OF ARBORICULTURE, A GUIDE TO THE PROFESSIONAL EVALUATION OF LANDSCAPE TREES, SPECIMEN SHRUBS, AND EVERGREENS (Urbana, Ill., 1975).

tures on their lots. Such laws will often assure collector owners that their neighbors cannot make structural additions that would shade the collectors. In highrise areas where all buildings are the maximum height, at least rooftop access is assured. But because existing buildings and land use controls were not designed with solar access in mind, there will often be much less sunlight available than is possible for any given density. Furthermore, additions, variances, and special exceptions can be made without consideration of solar access.

State legislation stating that local governments may zone for solar access is advisable.[12] But even without any amendment to state enabling legislation, courts would probably uphold the local use of the police power to regulate structures to protect solar access (see Chapter 4).

Shadows from Vegetation

Trees and other plants are vital to our survival. They contribute greatly to a pleasant, healthful urban environment. Although vegetation seldom serves as a direct source of energy in cities, properly placed trees and shrubs help to conserve energy by shading structures to reduce the need for air conditioning and by serving as windbreaks to reduce the need for heating. In many neighborhoods, however, shadows from trees will threaten solar collectors more than shadows from structures. In Washington, D.C., for example, there are 200,000 publicly owned trees and many more privately owned ones.[13] Much more research is needed on the extent and nature of shadows from vegetation and its effect on overall energy consumption before definitive recommendations for controls are possible. Controls will have to be adjusted to the particular needs of each community.

Vegetation controls may be more controversial in low-rise residential settings than are controls on structures. (In commercial and industrial areas, trees are generally not a problem; in high-rise areas, the restriction of development rights will be the more controversial issue.) This is because of the relative novelty of controls on privately

12. For example, Connecticut has amended its zoning enabling legislation to allow localities to adopt regulations that "encourage energy-efficient patterns of development, the use of solar and other renewable forms of energy, and energy conservation." Conn. Pub. Act No. 314 (1978) (to be codified as CONN. GEN. STAT. § 8-2).

13. MANAGING MUNICIPAL TREES, URBAN DATA SERVICE REPORT 8, no. 11 (November 1976): 2.

owned vegetation and because trees and large shrubs abound in residential areas and contribute to environmental amenities. The importance of trees has been summed up as follows:

> Nothing contributes as much to a pleasant city as trees. Besides their beauty and the shade they provide, trees and woodlands absorb pollutants such as carbon dioxide and sulfur dioxide; they help catch and settle particulates; and they help reduce noise levels.[14]

State and local laws and regulations rarely control the height and placement of vegetation on private property, although most cities do have some form of tree ordinance. References to vegetation control are rare in enabling acts, zoning ordinances, and other land use laws. In fact, case law on the subject suggests that there is no right to have vegetation on neighboring land removed as a nuisance even if it creates such dense shade that a house rots.[15]

The last section of this chapter takes a more detailed look at present vegetation control laws and speculates on the form new laws might take.

Correcting Common Misconceptions About Solar Access

Solar collectors are not necessarily on roofs. A common misapprehension is that solar collectors are always located on roofs. Collectors can also be located on walls and on the ground. A fairly common type of collector is in or on the south-facing wall.[16] Prefabricated walls that collect and store solar heat are available; they are typically used for warehouses. Even in already built-up high-rise areas, there may be a demand for south-wall exposure to sunlight. It cannot even be assumed that collectors will be attached to the structures they serve. Collectors may also be attached to accessory structures such as garages and fences, or may be ground-mounted.

Large structures, as well as small ones, may use solar energy. Another misconception is that solar energy systems will be useful only on small structures or single-family houses, and that solar access will therefore only be needed in suburban settings. On the con-

14. Michael McCloskey, *Nature and Cities,* SIERRA, April 1978, p. 16.

15. Granberry v. Jones, 216 S.W.2d 721 (Tenn. 1949).

16. Some communities may even want to protect the solar access of ordinary south-facing windows. It has been estimated that "inadvertent" direct thermal solar systems in California homes save about $20 million a year in fuel costs. David Bainbridge, *Planning for Solar Access: A Primer* (Sacramento: California Energy Commission, Solar Office, March 1978 draft), p.2.

trary, solar heating and cooling are already used in some large structures, including hotels, churches, laundries, fast-food establishments, warehouses, factories, and schools.[17]

Because structures in high-rise areas are often built to the same maximum height, it may be possible to guarantee rooftop access to all. In fact, tall buildings usually have flat roofs, so building orientation is less of a problem than with small structures.[18]

Retrofits for space heating are feasible. Solar energy systems are not limited to new construction, as many people believe. It can be economical to retrofit existing structures for solar space heating.[19] Existing shading and improper orientation will keep many homes and other buildings from using solar energy for space heating and cooling. But because hot water heating requires a smaller collector surface than space heating, less solar access would be needed.[20]

17. Examples are the eleven-story solar collector under construction on the Radison Hotel in St. Paul, Minnesota, and the 19,000 square feet of collectors built on top of Danbury Hospital in Connecticut to provide hot water. The large Frenchman's Reef Holiday Inn in St. Thomas, V.I., relies on solar collectors for air conditioning. *See* Frank Bridgers, *Solar Energy Applications to Large Buildings,* SOLAR AGE, February 1977, p. 10ff. Bridgers discusses situations where for large buildings a ratio of collector area to building area of only 10 percent is possible if heat pumps are used. (His article deals with space heating.) Albuquerque, New Mexico, has zoning to protect solar access in industrial parks and some manufacturing areas. ALBUQUERQUE, N.M., COMPREHENSIVE CITY ZONING CODE (1 May 1977).

18. The importance of structure orientation for solar access cannot be overemphasized. Jonathan Hammond, of Living Systems (Winter, California), says that even in northern latitudes and at densities of eight dwelling units per acre, good south-wall solar access is possible *if* structures are well-oriented.

19. *See generally,* SOLAR RETROFIT BIBLIOGRAPHY (Rockville, Md.: National Solar Heating and Cooling Information Center, 1978) (copies can be obtained by writing to P.O. Box 1607, Rockville, Md. 20850). In New Mexico and other parts of the Southwest, at least, it is apparently more common to retrofit existing homes for solar space heating than to incorporate it in new homes. One study credits these factors, among others: many homes in the Southwest have flat roofs and large yards, which made it easy to install and correctly orient collectors; the installed price of several types of solar equipment was the same whether it was to go on new or old homes; and the cost per Btu delivered by a solar system decreased as the size of a system increased—poorly insulated older homes enjoyed economies of scale. *See* Shaul Ben-David et al., *Near Term Prospects for Solar Energy: An Economic Analysis,* NATURAL RESOURCES JOURNAL 17 (April 1977): 202–05. These findings may or may not apply to other parts of the country.

20. A family of four that uses sixty gallons of hot water a day would require about forty square feet of efficient collector in a good orientation. BRUCE ANDERSON, THE SOLAR HOME BOOK (Harrisville, N.H.: Cheshire Books, 1976), p. 218. However, when estimating the need for solar access as a result of retrofitting, planners and

Specific Situations in Developed Areas Where Solar Access Is Unavailable or Uncertain

Even before determining exactly how much access to sunlight various collectors require, it is possible—in a very general way—to identify situations where solar access protection is currently inadequate or lacking and where owners of structures may have almost no assurance of continued access to the direct solar energy they now enjoy. In these situations minor amendments to present laws do not offer a satisfactory solution, as they do for new residential developments.[21] The descriptions of these situations are necessarily vague. There is no data base for a more precise analysis. And, as already mentioned, in some situations solar access protection is unnecessary. The following discussion describes a situation that ELI considers unprotectable under the police power, and several other situations that are protectable, but presently unprotected from shading by structures or vegetation.

Where Shading Is Already Prevalent

The following situation appears to be unprotectable (for economic reasons) on a comprehensive, areawide basis: where, on most lots in an area, all feasible collector sites are already shaded by existing structures, or will be shaded by construction that is underway or for which building permits have been issued. Similarly, where individual lots are already shaded (even in a neighborhood where solar access is generally available), solar access would be unprotectable on even a lot-by-lot basis for those shaded lots.

Although it may be legally possible for a state or local government to use the power of eminent domain in these situations (see Chapter 11), the drain on public funds would be so great that this option is effectively foreclosed. Under traditional police powers, the owner of an otherwise legal structure cannot be ordered to remove it.[22]

legislators should also keep in mind the possible widespread use of on-site photovoltaics in the future.

21. *See* American Planning Association, Protecting Solar Access for Residential Developments: A Guidebook for Planning Officials (Washington, D.C.: U.S. Government Printing Office, 1979). The ideas in this book also apply to large-scale urban redevelopment projects.

22. *But see* Robert M. Anderson, American Law of Zoning, 2d ed., 5 vols. (Rochester, N.Y.: Lawyers Co-operative Publishing Co., 1976), 1: §§ 6.64-6.73, pp. 505–32.

Areas that are heavily shaded by trees will sometimes be included in this unprotectable category because trees can be worth thousands of dollars. Although the police power could probably be used to require their removal, this would place a unique burden on their owners which would be unfair and politically unpopular. Where structures outlive trees, it would be possible to prohibit the replacement of certain trees. Another possibility would be to require specified types of trees removed and to require the benefitted person to pay for the cost of removal and the tree owner's loss of property value. Trees that save more energy than could be had by their removal should be exempted.

In an absolute sense, no situation is entirely unprotectable. If an important enough public purpose exists, troublesome structures and trees could be declared nonconforming uses and be amortized over a reasonable period. It is doubtful, however, that the public presently feels the sense of urgency about solar energy that would make such a law acceptable.

Where Changes in Present Land Use Controls May Result in the Shading of Otherwise Secure Solar Collectors

Even when present laws regulate the size of lots and the height and placement of structures in a manner that provides good solar access, the law may change. For example, imagine that the owner of a lot in an area zoned for large lots and one-story houses puts a collector where it will not be shaded by any development allowed on neighbors' land. If the city wishes to rezone the land immediately to the south for five-story apartment buildings, it usually may do so. The fact that someone has installed a collector does not protect against a rezoning, since one cannot acquire a vested right in a particular zoning.[23] Areas may not be rezoned on a whim, of course.[24] Similar shading problems could arise, without zoning changes, from the granting of variances.

23. In Reichelderfer v. Quinn, 287 U.S. 315, 323 (1932), the Supreme Court said "zoning regulations are not contracts by the government and may be modified." *See also* Zopfi v. City of Wilmington, 273 N.C. 430, 160 S.E.2d 325 (1969); Robinson v. City of Los Angeles, 146 Cal. App. 2d 810, 304 P.2d 814 (1956).

24. *See* Baker v. City of Algonac, 39 Mich. App. 526, 198 N.W.2d 13 (1972); Fasano v. Board of County Comm'rs, 264 Or. 574, 507 P.2d 23 (1973); Cain v. American Nat'l Bank & Trust Co. of Chicago, 26 Ill. App. 3d 574, 325 N.E.2d 799 (1975).

Where Shading Would Occur if Structures Were Extended Up to Present Zoning Limits

In many urban and suburban areas where houses are only one story, zoning allows two or three stories. This situation is deceptive because it suggests that more restrictions on density exist than are actually on the books. It is unwise to assume, on the sole basis of existing solar access, that present laws adequately protect solar system owners. The expansion of structures—whether upward or outward—may result in shadows on collectors when homes are close together. South-wall and ground collectors will be more vulnerable than those on roofs. Underdevelopment also occurs in commercial and industrial areas, where development rights may be much more valuable than in residential zones, and where rising property taxes encourage full development.

Where a Zone Allows Higher Densities than a Zone to Its North, East, or West

Zone boundaries often run down the middle of streets (rather than splitting a block into two different zones). In some cases, the streets are wide enough to mitigate shading problems. But in areas with tall buildings compared to street widths, shading across zone boundaries may have to be reckoned with. A problem may exist, for example, in the blocks near and to the north, east, or west of a central business district. Where areas are already fully developed, there is little that can be done, except on a long-range basis as individual buildings are replaced. But where districts are in transition or where the one to the south, west, or east is still underdeveloped, it is possible to impose "step down" zones, which permit gradual changes in height, from block to block, reducing the effect of shadows. Such changes should only be made, of course, where they are consistent with the goals of an overall plan for the city.

Where Public Construction Could Shade Sites

Like other property owners, governments may and do develop space above their land. Occasionally, however, people owning land adjacent to public property can acquire an easement for light. All such cases involve light crossing public transportation routes or viaducts.[25] These easement cases are the rare exception to the general

25. Compensation was paid to landowners in New York who had relied on the formerly open space for light. The first precedents were the famous Elevated Railroad cases, Story v. New York Elevated R.R. Co., 90 N.Y. 122 (1882). A long series of

rule that property owners in the United States have no right to light crossing nearby land. There is no equivalent right to light and air across privately owned property.[26]

Where Vegetation Is Not Regulated

Vegetation can also be expected to shade collectors. For instance, the most frequent source of conflicts over shadows from public lands will probably be trees in parking strips along streets. In many communities such trees are publicly owned and maintained. Even if land-use regulations that include controls on the height of structures and vegetation are imposed on private landowners in a zone, local governments are not necessarily bound by these rules, depending on the wording of the ordinance and judicial interpretation of it. (Nor are the state and federal governments bound by such regulations.)

Because private vegetation is largely unregulated, the owners of structures that are shorter than surrounding trees will be at maturity may not be assured of even rooftop access to solar energy (depending on how tall and how far away the trees are). Under common law, property owners may have a right to prune any branches extending over their lots, but because sunlight does not come down vertically, this protection is not enough.

Special Vegetation Control Problems

Because of the novelty and importance of controlling urban vegetation, additional information follows on the limitations of present laws, the mechanics of tree care, the value of trees, and the directions that new laws might take.

subsequent cases, referred to as "one of the most important and interesting chapters in the history of litigation," Powers v. Manhattan R.R. Co., 120 N.Y. 178, 183, 24 N.E. 295, 296 (1890), culminated in Muhlker v. New York and Harlem R.R. Co., 197 U.S. 544 (1905), in which the Supreme Court affirmed the New York decisions. *Muhlker* was upheld in Roberts v. New York City, 295 U.S. 264 (1935). Iowa courts have also required compensation. In Liddick v. City of Council Bluffs, 232 Iowa 197, 5 N.W.2d 361 (1942), a person owning land abutting a public street where the city proposed to construct an elevated viaduct won compensation. The court said that the lot and the easement were used and enjoyed as a unit and that the substantial impairment was a taking. The holding was reaffirmed and extended to property abutting a state highway in Anderlink v. Iowa State Highway Comm'n, 240 Iowa 919, 38 N.W.2d 605 (1949), and in Rhodes v. Iowa State Highway Comm'n, 250 Iowa 419, 94 N.W.2d 97 (1959).

26. Marchant v. Pennsylvania Co., 153 U.S. 380 (1894).

Present Laws and Precedents Are Inadequate

Several means of protecting solar access are described in the following chapters. Whatever controls are adopted should apply to both public and private vegetation. In general ELI considers it preferable to use a single coordinated set of controls over both structures and vegetation, although it would be possible, for example, to regulate vegetation through nuisance law and structures through zoning. On the other hand, there is no legal reason why different types of solar access controls could not be placed on buildings or on vegetation.

Existing municipal laws that regulate vegetation are neither good models nor strong precedents for solar access legislation. They serve a variety of limited purposes:[27]

- ☐ To give a city the power to regulate street trees
- ☐ To allocate liabilities related to street trees
- ☐ To establish protective standards for municipal (and possibly private) trees
- ☐ To regulate private trees that pose a hazard or nuisance to persons or property outside the boundaries of their lots[28]

Only 11 percent of the cities surveyed in a Michigan State University study reported a law that restricts the cutting of privately owned trees.[29] Thus, most people who want to remove trees on their own lot to gain more sunshine for themselves or their neighbors should be able to do so. (Restrictive covenants may prevent private tree cutting in some subdivisions.) In addition to these types of laws,

27. MANAGING MUNICIPAL TREES, p. 11.

28. An example of this last type of law is the firebreak provision in LOS ANGELES MUNICIPAL CODE § 57.25.20 (1976):

> All native brush situated within one hundred feet of any structure . . . shall be maintained at a height of not more than eighteen inches above the ground nor less than three inches above the ground and shall have removed therefrom all branching growth and leafy foliage.

Exceptions are made in the case of specimen native shrubs that do not exceed 216 cubic feet in apparent volume and that are adequately spaced from one another and from structures. These provisions are interesting as an example of how stringent vegetation controls may be, and how heavy a burden they may impose on property owners, when they are justified by an important public purpose.

29. MANAGING MUNICIPAL TREES, p. 11. For example, chapter 104 of the FAIRFAX COUNTY, VIRGINIA, CODE requires landowners to obtain permits if they plan to clear over 5,000 square feet of land.

there are weed control ordinances, laws forbidding the removal of certain vegetation needed for soil erosion control, and ordinances that limit the height of vegetation that could block motorists' views of traffic conditions.[30] In the following cases concerning the use of the police power to control vegetation, courts upheld that use, at least in part. Whether the police power could be applied in solar access cases, however, is unclear because the controls needed to provide solar access would be more restrictive. In *Stevens v. City of Salisbury,*[31] a Maryland court upheld sections of an ordinance limiting the height of hedges and shrubs near intersections to three feet. However, the court specifically rejected the section that would have required landowners to remove "all barriers except buildings and tree trunks above the established curb location."[32] This provision was said to be unreasonable because it was not limited to specific types of trees that could be trimmed without losing their value. Furthermore, the court continued, it is unreasonable to make property owners destroy or substantially reduce their property's value at their own expense.[33]

A well-known U.S. Supreme Court case, *Miller v. Schoene,* held that diseased red cedar trees could be destroyed because they threatened apple trees.[34] It may possibly be a more useful precedent for solar access controls. Although the opinion states that the police power can be used to require the destruction of vegetation when a legitimate public interest is involved, the factual setting was very different from that in a conflict over solar access.[35]

30. *See,* e.g., Derby, Connecticut, Zoning Ordinance 8-D-50 (1969), which limits trees and shrubs to certain areas of corner lots to three feet in height (or requires that tree branches be trimmed away to a height of at least ten feet above curb level). The purpose is traffic safety.

31. 240 Md. 556, 214 A.2d 775 (Ct. App. 1965).

32. 214 A.2d at 783–84.

33. Ibid. *See also* People v. Berlin, 62 Mis. 2d 272, 307 N.Y.S.2d 96 (1970), upholding an ordinance restricting hedges along rear property lines to six feet in height. This court said the ordinance was authorized by section 1 of the Standard State Zoning Enabling Act.

34. 276 U.S. 272 (1927).

35. A single cedar tree that hosts cedar rust can infect apple trees owned by many persons. Something harmful was emanating (trespassing, in a sense) from one property owner's land to the land of others. In the case of shadows, it is the *absence* of light that generates conflicts. Several reviewers are more optimistic about the usefulness of this case than ELI is. They note a similar strong public purpose favoring regulation to protect solar access, and feel the *presence* of shade is analogous to the presence of rust.

Value, Cost of Care, and Choice of Trees

The data that exist on urban trees deal almost exclusively with publicly owned trees, such as those planted along streets.[36] Although larger cities tend to have more street trees per linear mile, the mean for all cities is around 69 trees.[37] The average value of these trees is between $300 and $400.[38] Privately owned trees have even greater value and often increase the value of a home by 5 to 10 percent.[39] The monetary value of trees is one reason that this study does not recommend laws that require the destruction of existing large trees on lots adjacent to collectors.

The cost to the public of pruning trees should also be considered when solar access laws are drafted. Caring for public trees now takes only a small part (0.56 percent) of a typical city's budget[40] and is viewed by many cities as a nonessential activity.[41] If much more extensive pruning or replacement of public trees is required for solar access, financially pinched cities could require the owner of a solar collector to pay for the trimming of public trees that grow to shade the collector.

Most cities reportedly allow homeowners to plant trees in the city right-of-way adjacent to their property. Cities typically require advance approval and thereafter regulate the trees but a significant number do not.[42] To protect solar access, ordinances should be amended to prohibit either municipalities or private persons from planting new street trees in locations where they shade collectors. By

36. The material in this section is based largely on a survey conducted in 1974–1975 by the Department of Forestry, Michigan State University. Questionnaires were sent to 1,426 municipalities; 864 returned with useful data. The survey is discussed in MANAGING MUNICIPAL TREES.

37. Ibid., p. 3.

38. Ibid., p. 4.

39. BRIAN R. PAYNE, TREES COULD MAKE THE DIFFERENCE IN THE SELLING PRICE OF YOUR HOME, Forestry Science Photo Story no. 26 (Upper Darby, Pa.: U.S, Department of Agriculture, Northeastern Forest Experiment Station, 1975). The value of individual trees on private property has been estimated to be between $500 and $1,500. MANAGING MUNICIPAL TREES, p. 4.

40. Ibid.

41. Ibid., p. 6. Most tree maintenance is done by city agencies (44 percent delegate this work to the parks and recreation departments, about 10 percent have a special forestry or tree care division, and small cities tend to use the street or public works department) (ibid., p. 4). Many cities also hire private firms to undertake some of the tree care. Larger cities are more likely to employ a professional forester or arborist.

42. Ibid., p. 11.

planting appropriate species, it will often be possible to enjoy both trees and solar access. For instance, where solar energy is used only for space heating in the winter, deciduous trees with nonpersistent foliage (those that shed leaves in the autumn) and with minimal shading from branches may not pose a serious problem for collectors. However, even the bare branches of most species of deciduous shade trees would impair the functioning of a solar collector in winter.[43]

Grandfathering

ELI believes that trees that shade collector sites or that violate solar height regulations at the time a solar access control becomes effective should generally be allowed to remain.[44] An analogous situation arises when a structure does not meet the requirements of a new zoning law. The building is said to be a nonconforming use. Typically, such structures are allowed to remain indefinitely or for a certain period of time; they are said to be "grandfathered" or "timezoned." Although laws have been upheld that require nonconforming uses to come into conformity within a reasonable time,[45] in the cases of trees this may not be practical.

A separate question is whether future growth of existing trees should also be grandfathered. This is more difficult to answer. There are practical problems: How does one prove how large the tree was at the time the ordinance was enacted (or the collector installed)? Who has the burden of proof? Who bears the cost of pruning? Furthermore, the pruning costs may be very high when a tree is tall and located far from the street. (Street trees are often pruned from trucks that roll down the street.) A related question springs from the wording of several proposed statutes that exempt vegetation from regulation if it "shades a collector at the time of installation." What if the

43. For more discussion of this, *see* APA, Guidebook for Planning Officials, and idem, Site Planning for Solar Access: A Manual for Developers and Site Planners (Washington, D.C.: U.S. Government Printing Office, 1979). These books contain information on shading by bare twigs and branches for some species for one part of the country. Unfortunately, these factors differ greatly with geographic location. Information of this sort is relevant to existing developments whenever new plantings are contemplated.

44. Two reviewers questioned this statement. They suggested that pruning may be worth requiring, or that a city should consider requiring the cutting of trees that shade a collector when it is installed.

45. *See* City of Los Angeles v. Gage, 127 Cal. App. 2d 442, 274 P.2d 34 (1954); E. B. Ellicott Adv. Co. v. Metropolitan Dade County, 425 F.2d 1141 (5th Cir.), *cert. denied,* 400 U.S. 805 (1970).

collector is only a little shaded? May the tree then grow to shade it completely? Careful drafting should make the intent of the government clear.

If the law states that vegetation is exempt if it existed (or pierced a solar envelope) at the time of enactment (or installation of a collector) or that such vegetation is exempt but only to the extent of its existing growth, then problems of proof arise. Several forms of proof are possible. Photographs, for instance, could resolve many disputes. They could be aerial photographs of an entire community taken on the relevant date. The height of trees could be calculated from their shadows in the photographs. Seedlings and other small plants may not show up on such photos, but the law could state that trees or bushes too small to show up would not be grandfathered. Aerial photos would be impractical if the relevant date varied—if it depended, for instance, on when individual collectors were installed. In such cases ordinary photographs of the yard and trees involved, appropriately dated and notarized, should suffice. However, a deciduous tree that does not appear to shade a collector in the winter may shade it in the summer. Regulations should deal with such possibilities. Another option is to register trees. In individual cases this might be feasible, but registering all trees and shrubs on a community-wide basis is not a realistic option.

Who should be responsible for proving that the vegetation existed at the relevant time—the collector owner or the vegetation owner? It seems logical to put the burden on the party with the greatest self-interest in preserving the vegetation—the owner of the tree or shrub. The burden is not a heavy one. The cost of a photograph plus notary fees should be only a few dollars. Information on what to include in a photograph could easily be given property owners at the time they are notified of a potential collector installation, or at the time solar envelope zoning becomes effective. Another reason that the burden should be on the vegetation owner rather than on the solar user is that the solar user would have to prove something did *not* exist, which is more difficult than proving something does exist.

2
Solar Access Basics

Discussions of the law of solar access, as well as proposed and enacted solar access legislation, have tended to gloss over the technical requirements of solar energy collection. Some basic knowledge of collectors, sunlight, and shadows is crucial to designing and implementing solar access laws. While these technical concerns need not dictate the legal means for solar access protection, they must be taken into account if solar access is to be adequate, yet reasonable, in terms of the restraints it places on nearby lands.

Appendix C describes the basics of two common solar applications, residential solar heating and domestic hot water. Appendix A contains detailed discussions of some technical aspects of describing solar position and shadows.

Kinds of Solar Energy

Solar energy includes direct, diffuse, and reflected components. Because direct solar energy is generally the major component and is most subject to shadows caused by structures and vegetation, it is of the most interest to those writing legislation to protect solar access. Thus, when the words sunlight or sunshine are used in this report, they refer to direct solar energy. If there is reasonable access to direct solar energy, diffuse radiation (which comes from the whole sky but is most intense near the sun) need not be separately protected.

Solar energy can also be reflected from bodies of water, trees, and structures near collectors, but most reflected solar energy comes from the ground.[1] In some low density areas, access protection could be extended to the ground below and to the south of collectors to take advantage of reflected energy. In high density zones it would be impractical to protect this extra access. (This can be treated as an issue distinct from whether to secure access to sunlight for constructed reflectors[2] or yard collectors—those not located on buildings.)

Applications of Solar Energy

The principal current application of solar energy to structures is for residential water and space heating; the next most common application is nonresidential water and space heating. Other applications are in more experimental stages: process heating for commercial, industrial, and agricultural uses; solar-induced draft ventilation and absorption cooling driven by solar heat; electricity generation from solar heat; direct generation of electricity from sunlight (photovoltaics); mechanical energy from solar-driven heat engines; and chemical energy from photolysis and photosynthesis. The use of sunlight for illumination has been the subject of recent experiments on ways to collect light more effectively than windows can and piping light to inner rooms that receive no natural lighting.[3] Some of these applications may become widespread within a decade and may then require solar access protection.

It might be assumed that these different kinds of solar energy applications need different kinds of solar access. Yet it appears that the common fundamental need is for as much solar access as can be economically (and politically) justified. The more marginal an application is economically, the more access it will need to be cost effective, but the same solar access principles—both technical and legal—

1. A typical ground reflectivity of 20 percent is used by the American Society of Heating, Refrigerating, and Air-conditioning Engineers (ASHRAE). ASHRAE, Handbook of Fundamentals (New York, 1972), pp. 388–92. This figure is clearly low for northern latitudes in winter, where fresh snow can reflect 75 percent of incident sunlight. B. D. Hunn and D. O. Calafell II, *Determination of Average Ground Reflectivity and Solar Collectors,* Solar Energy 19, no. 1 (1977): 88. On the other hand, if sunlight does not strike the ground below and to the south of a collector, the reflected energy may be negligible even if there is snow because there is little light to be reflected.

2. *See* Steven C. Seitel, *Collector Performance Enhancement with Flat Reflectors,* Solar Energy 17, no. 5 (1975): 291.

3. *Sun Lit Fuel Saver,* Solar Age 3, no. 5 (May 1978): 10; *Sandia Solar Lighting Proves Economical,* Solar Utilization News, February 1977, p. 3.

that apply to current applications should also be appropriate to more exotic future applications.

Types of Solar Collectors

Solar energy systems are often described as "active" or "passive." As only the collector surface portion of a solar energy system requires access to sunlight, it is immaterial to this study what occurs inside a collector, or whether a solar system is active or passive. Passive systems tend to use vertical south-wall collector surfaces (such as windows) and active systems tend to use rooftop collectors, but there are myriad exceptions to this.[4]

Even where south-facing walls are frequently shaded, such as in urban environments, passive systems merit consideration as candidates for solar access protection. A passive system can perform comparably to an active system for the same area of collection surfaces.[5] Furthermore, there is some evidence that passive systems tend to cost less than active systems, both in new structures and when existing structures are retrofitted for solar energy use. For these reasons, vertical south-wall collectors deserve access protection whenever possible.

Tracking or fixed collectors constitute another broad classification. Tracking collectors move to follow the position of the sun to collect the most direct energy available. Fixed collectors do not move but can be in many different positions relative to the structures they serve. Either tracking or fixed collectors may be used for most solar applications. Tracking collectors are not yet widely used but may eventually require separate legislative consideration because they are more sensitive to early morning and late afternoon shadows from the east and west than are fixed collectors. Under some circumstances this may justify protecting them from shading over a much wider area and for longer hours than is necessary for other collectors.[6] Examples might be where access is readily available at little cost, or where tracking collectors are expected to be popular.

4. *See* the illustrations in Appendix C for examples of roof-mounted passive collector designs. Active collectors may be south-wall-mounted or ground-mounted in some designs. *See,* e.g., TOTAL ENVIRONMENTAL ACTION, SOLAR ENERGY HOME DESIGN IN FOUR CLIMATES (Church Hill, N.H., May 1975), p. 96, for a vertical south-wall-mounted air collector design for Minneapolis, Minnesota.

5. J. D. Balcomb, J. C. Hedstrom, and R. D. McFarland, *Simulation Analysis of Passive Solar Heated Buildings: Preliminary Results,* SOLAR ENERGY 19, no. 3 (1977): 277.

6. For perhaps as much as two hours a day more. *See* Melvin M. Eisenstadt,

ELI generally recommends against a community's tailoring solar access protection to the needs of specific solar technologies, however, because it is not yet known which types of solar energy systems and collectors will prove most effective under different conditions. Laws that favor specific technologies could dictate choices that may later be proven unwise.

Collector Locations

It has been suggested that only rooftop collectors or those at least ten feet above the highest floor merit access protection.[7] Such an assumption can lead to inadequate solar access laws because there are many other valid locations for collectors. As noted, most passive designs do not have collectors on rooftops. And in many buildings, such as Cape Cod cottages, roofs may begin less than ten feet above the top floor so that not even all the roof would be shadow-free under a law that only protected collectors at least ten feet above the highest floor.

Four broad options concerning degree of solar access protection can be listed:

- Only part of roof
- All of roof
- South wall and roof
- Part of south yard (if any) plus south wall and roof

These different options allow for greater or lesser protection depending on local conditions (which may vary from area to area within a city), local policies concerning desirable densities, and the desirability of using the various applications of solar energy systems. The range of options for protected locations is actually broader than this list suggests, however. This is especially true if more than single-family residential use of solar energy is contemplated. Figure 2-1 shows some of the major variations possible. The collector sites illustrated are shown in increasing order of restraints placed on

Stephen C. M. Long, and Albert E. Utton, *A Proposed Solar Zoning Ordinance,* mimeographed, prepared for the New Mexico Energy Institute (ERB-76-103, November 1977), p. A-19. Revised version, same title, URBAN LAW ANNUAL 15 (1978), p. 247.

7. Melvin M. Eisenstadt and Albert E. Utton, *Solar Rights and Their Effect on Solar Heating and Cooling,* NATURAL RESOURCES JOURNAL 16, no. 2 (1976): 390–91.

(1) Northly part of flat roof or upper part of south-facing slope of roof.

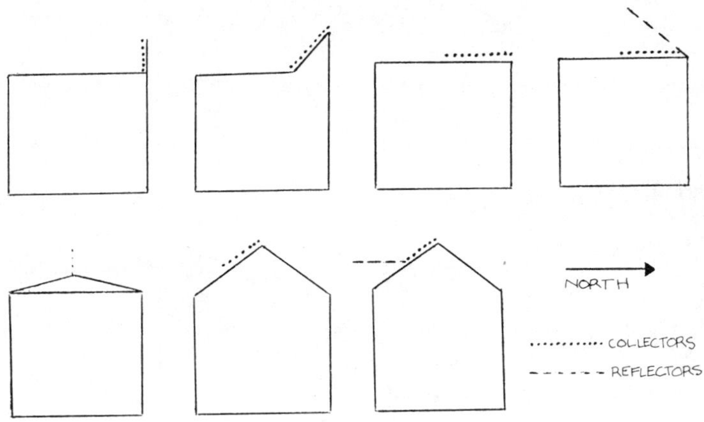

(2) All of roof or all of south-facing slope of roof.*

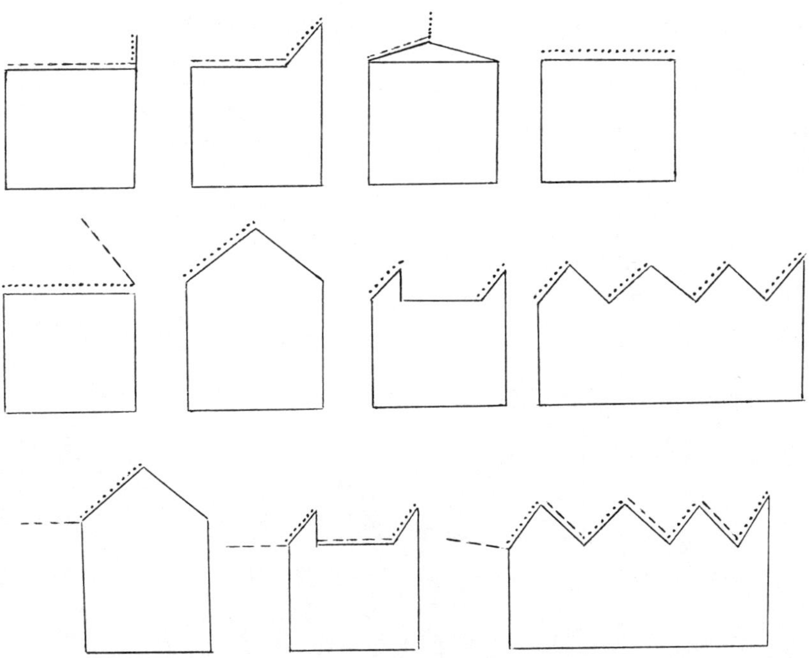

FIGURE 2-1
EXAMPLES OF COLLECTOR LOCATIONS IN RELATION TO STRUCTURES SERVED

(3) Upper part of south-facing walls.*

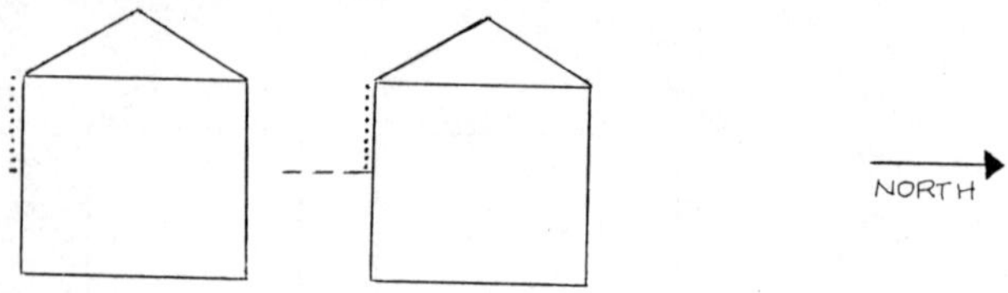

(4) All of south-facing walls.*

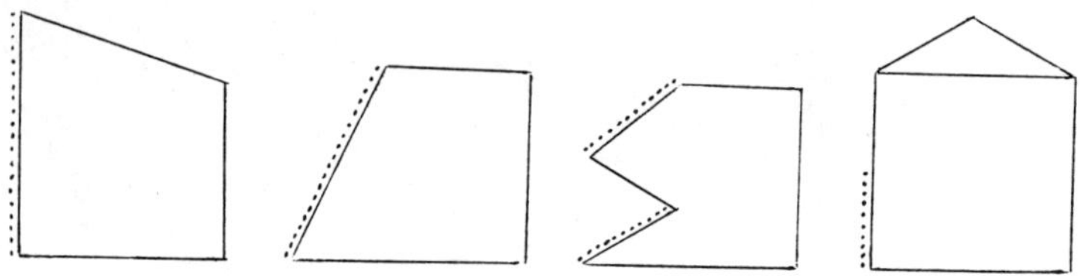

(5) Ground just to south of walls.*

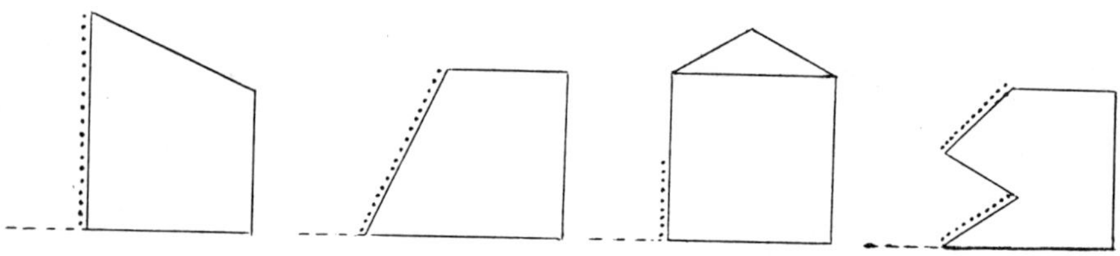

(6) Any part of lot meeting stated setback and height standards.*

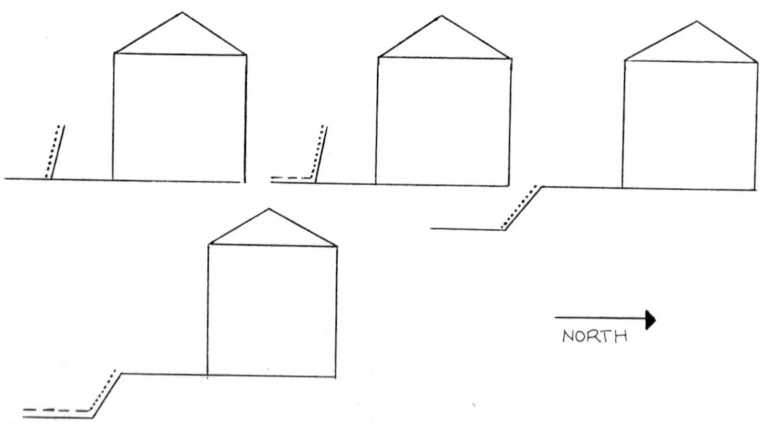

(7) All of the lot (not shown).*

* Each option includes all of the options above it in the figure.

FIGURE 2-1 (continued)

neighboring properties when the same fraction of available energy is protected.

Sites could be protected from shading:

- □ Only for collectors, as defined
- □ For collectors and reflectors associated with them as defined (such as mirrors, metal foil, and possibly white rock)
- □ For collectors, reflectors, and natural reflectors that are part of a collector design (such as a body of water, sand, grass, snow)
- □ For whatever may be there, whether a wall, patio, garden, or yard (on the assumption that it is a likely site for future collectors)

Each locality must determine which sites are and will be commonly used and which of these can be protected without unreasonable restraints on development.

Movements of the Sun and Shadows

Movements of the Sun

The apparent movements of the sun follow a predictable, repeating pattern.[8] The pattern can be described in words, pictures, graphs, tables, and mathematical notation. Although the last way is most accurate, it is also hardest to understand. The description here uses the first four methods.[9]

The apparent location of the sun at any particular time depends on three factors: the observer's location (latitude),[10] the time of day (solar time), and the time of year (date).[11] Figure 2-2 sketches the general nature of the sun's path at the solstices and equinoxes.[12]

8. Although the "movements of the sun" are an illusion that results from the earth's daily rotation on its axis and annual orbit around the sun, it is usual and useful to think in terms of movements of the sun relative to wherever we are or to the point we are thinking about. Our everyday expressions such as "the sun rises and sets" also take this viewpoint.

9. Appendix A's discussion of the movements of the sun also gives the formulas for them.

10. The distance, measured in degrees, between the equator and the location in question.

11. The date determines the declination or angle between the equator and the sun; this angle varies from roughly −23.5 degrees at the winter solstice (typically December 21) to +23.5 degrees at the summer solstice (typically June 21), and is 0 degrees at the equinoxes (typically March 21 and September 21).

12. Figure 2-2 and the discussion in Appendix A apply to the Northern Hemisphere only.

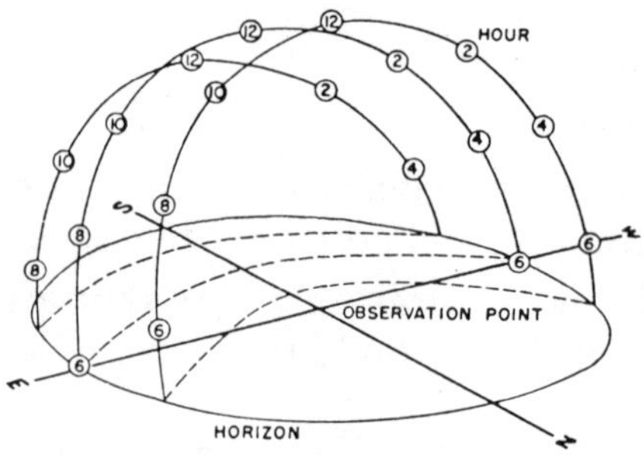

78. Half sphere of the imaginary sky-vault with the sun-paths. The projections of the sun-paths are shown in dashed lines on the horizon plane.

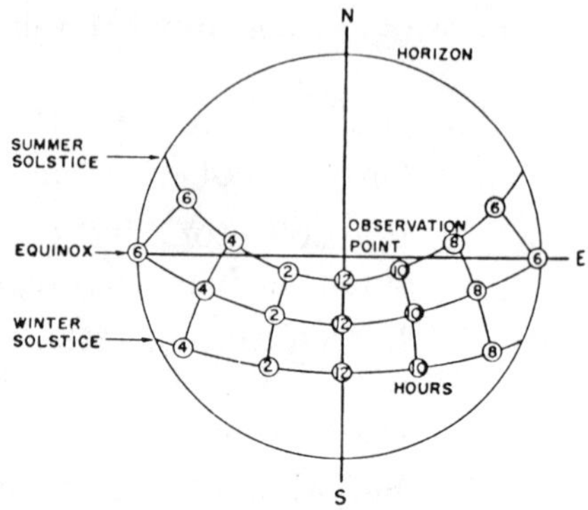

79. Projected diagram of the sky-vault, called "sun-path diagram."

From Figures 78 and 79 in Aladar and Victor Olgyay, *Solar Control and Shading Devices* (copyright © 1957 by Princeton University Press; Princeton Paperback, 1976), p. 38. Reprinted by permission of Princeton University Press.

FIGURE 2-2
SEASONAL VARIATIONS IN THE SUN'S DAILY PATH ACROSS THE SKY

Shadows and Shading

While the position of the sun determines shadow directions and lengths, it is the shadows themselves that are crucial to solar access and solar access laws. The length of a shadow cast on bare ground by a vertical pole, for example, depends on the altitude of the sun, height of the pole, and slope of the ground in the direction of the shadow (see Figure 2-3). The lower the altitude of the sun, the longer the shadow is. The taller the pole, the longer the shadow is. The more steeply the ground slopes downward (or upward) in the direction of the shadow, the longer (or shorter) the shadow is, up to the point where the ground slopes down at the same or a greater angle than the sun's altitude, when the shadow is infinite—that is, the sunlight never reaches the ground after passing over the pole.

The effect of slope on total shadow length at the equinoxes and summer solstice is much less severe than at the winter solstice. Thus, the winter solstice shadows appear to be the limiting factor on nearby development if solar access is to be protected (see Appendix A).

Both buildings and vegetation, of course, may cast shadows. In many areas, vegetation acts as a summer cooling agent; the placement of trees for this purpose may sometimes interfere with winter solar energy needs. In new communities, vegetation could be sited to provide maximum solar efficiency over the whole year.

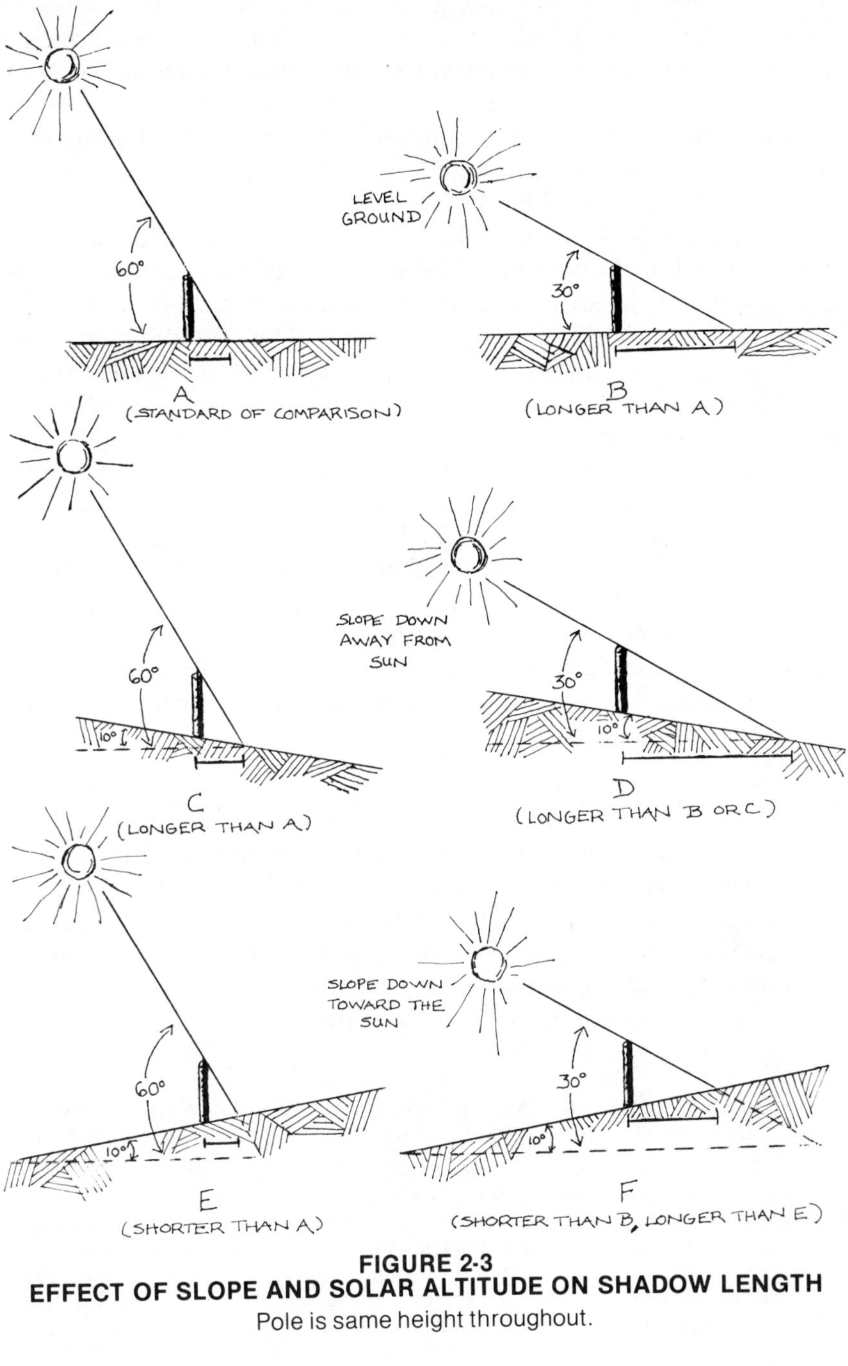

FIGURE 2-3
EFFECT OF SLOPE AND SOLAR ALTITUDE ON SHADOW LENGTH
Pole is same height throughout.

In developed areas, existing vegetation may present severe solar access problems. Vegetation may leave only some rooftop access or, if trees are tall and wide, none worth protecting. Neighborhoods that have few large trees should adopt vegetation controls to preserve solar access unless there is a countervailing need for shading (as in the Southeast where summer cooling problems are often more severe than winter heating problems).

Appendix A shows one way to evaluate the solar access for an existing neighborhood where the sizes, locations, orientations and spacings of all the structures and trees are known. (This could also be used for a planned new development.)[13] The example used there is a simple, but not implausible, situation. The same formulas could be applied to other, more complex situations, with structures of varying heights or with sloping sites.

The key to good solar access protection is to know what kind of development exists. Solar access can then be calculated[14] for what already exists and for what lawfully could be built. Alternative setbacks, height limitations, and other controls could be considered (where they would not create nonconforming uses of many existing structures and vegetation), and a decision made concerning what access is to be preserved and at what price. The price might be less intensive development or money out-of-pocket for a private arrangement (such as an easement).

Specification of Protected Solar Access

The extent of protected solar access can be stated in terms of time or space; that is, shading can be prohibited if it results from incursions into specified spaces or if it covers a collector surface during specified times. Some forms of these two options offer more advantages for solar access protection than do others. Laws may define solar access protection in terms of both time and space.

13. *See* American Planning Association, Protecting Solar Access for Residential Developments: A Guidebook for Planning Officials (Washington, D.C.: U.S. Government Printing Office, 1979), for a different approach.

14. ELI is now working on methods for doing the calculations quickly on programmable pocket calculators (which cost about two hundred dollars) or more slowly on inexpensive scientific pocket calculators (which cost about twenty dollars). Nevertheless, the hardest part of solar access evaluation will continue to be data collection concerning what structures and vegetation exist in typical situations within the various areas of a community.

Time Options

Standard time. The most common specification of protected access in proposed legislation (and easements and restrictive covenants) is that collectors shall not be shaded between the hours of, say, 10 A.M. and 2 P.M. or 9 A.M. and 3 P.M. or 8 A.M. and 4 P.M. local standard time.[15] The use of standard time has the obvious advantage of simplicity and familiarity. In both state and local legislation, standard time would be a single uniform measurement.[16] Standard time (except during daylight-saving time) is the familiar time on everyone's clock.

But the sun operates on its own time, which differs from our standard twenty-four-hour day. If standard time were used, there would be two effects: the effect of local longitude and the effect of the "equation of time."

Reliance on standard time to protect solar access can result in less favorable solar access and the placing of unreasonably unequal burdens on neighbors to the east and west.[17] This is because of the effect of local longitude on shadow lengths within each standard time zone at any given time. For example, at 3 P.M. standard time within a single time zone, shadows will be longer in the eastern part of the zone and shorter in the western part, compared to the middle of the time zone.

Compensated standard time. Because of the problems with using standard time, the effect of local longitude should be taken into account: a simple alternative to the usual symmetrical hours of standard time is to use standard time with a correction for local longitude. Table 2-1 shows how standard time hours could be adjusted for local longitudes in a single time zone, yielding protection from 9 A.M. to 3 P.M. solar time. (In general, compensation for longitude can be made by subtracting four minutes per degree for localities west, and adding four minutes per degree for localities east, of the standard meridian for the time zone.)

15. E.g., Colo. S.B. 38 (1976); N.M. H.B. 325 (1977).

16. Or so it appears. A recently enacted California law, for instance, specifies 10:00 P.M. to 2:00 P.M. standard time statewide. Cal. A.B. 2321, § 1 (to be codified as CAL. PUB. RES. CODE § 25982) (passed 31 August 1978). At the western edge of the state, this is 9:43 A.M. to 1:43 P.M. solar time, and at the eastern edge, this is 10:25 A.M. to 2:25 P.M. solar time at the winter solstice. This variation with local longitude hardly makes sense, as shown in the discussion in Appendix A of the effect of using standard time.

17. *See* the discussion of this in Appendix A.

TABLE 2-1
STANDARD TIME COMPENSATED FOR LONGITUDE

Standard Time Hours	Relationship to Standard Meridian
9:30 A.M.–3:30 P.M.	7½°W
9:15 A.M.–3:15 P.M.	3¾°W
9:00 A.M.–3:00 P.M.	0°
8:45 A.M.–2:45 P.M.	3¾°E
8:30 A.M.–2:30 P.M.	7½°E

The "equation of time" is another correction to standard time (figured from solar time) due to the variations in the length of the solar day compared to the average twenty-four-hour day. During the course of a year, the equation of time varies over a range of about half an hour. When the effect of the equation of time on the use of standard time (compensated for local longitude) is measured, the variation in shadow lengths is not profound. A final slight compensation (adding 1.6 minutes to hours compensated for longitude) could be made. This is not essential, but if standard time is to be compensated at all, this second step is hardly any more trouble.

It is difficult to see any serious disadvantages to these adjustments to standard time hours. They do require figuring a single number correction factor for any particular locality or for each longitude within a state for statewide laws.[18] But the factor does not need to vary with time of year and could be easily understood by everyone. At the same time, local weather peculiarities, such as early morning or late afternoon fog, should be taken into account. These can be dealt with in the same way by adjusting hours to suit the locality.

Solar time. Under solar time "noon" always occurs when the sun is due south. It is possible to specify solar time standards in a law or agreement protecting solar access. Earlier drafts of this study recommended that course. ELI is now convinced, however, that if standard time is compensated as described above, there is not enough difference in the resulting protected solar access or burdens on nearby properties to justify the use of an unfamiliar time scheme. But solar time with all its complexities is still better than uncompensated standard time (except near the standard meridians).

18. For localities close to the standard meridian for the time zone, the effect of the tentative hours to be protected could be calculated using the formulas in Appendix A, and a decision could be made as to whether this small complication is worth the trouble. For some localities near the standard meridian, the use of compensated standard time would not be worthwhile, but for most localities it would be, especially for localities near time zone boundaries.

Space Options

Although most solar access bills and legislation have been written in terms of specified hours of protection, some easement bills and legislation require solar easements to be written in terms of "angles expressed in degrees" rather than hours.[19] (Whether an easement written in terms of hours would be valid and could be registered in the land title registry, where such a law has been passed, is discussed in Chapter 11.) There are two broad approaches to defining solar access in terms of space:

- □ In relation to a collector site on the land of the benefited party
- □ In relation to the lot lines of the land of the burdened party

Appendix A has a detailed explanation of how planes can be defined for solar access protection under both of these approaches.

It should be remembered that what is reasonable under one set of circumstances for a given level of protected access should not be assumed to be reasonable under different circumstances. Thus, it is best to evaluate available access on an area-by-area basis and provide rules appropriate to each area. This evaluation must consider whether the amount of solar access sought is consistent with other community goals.

Time/Space Hybrid Options

Another way of specifying solar access is to prohibit shading of a specified space during specified times. In a sense, this amounts to regulating locations of collectors by affording protection during particular hours only for collectors located as specified.

Yet another is the hypothetical wall method. Under this method it is forbidden to shade more area than would be shaded by a hypothetical wall around the perimeter of the protected property during particular hours.[20] The protection available under this method can be seen by figuring the implied plane of protection, found by the method used in Appendix A. (Appendix A has a hypothetical wall example worked out to show the implied plane of protection.) Such

19. E.g., 1975 Colo. Sess. Laws ch. 326; Fla. H.B. 3087 (1976); Ill. H.B. 1512 (1977).

20. This method is adopted in Los Alamos County, N.M., Zoning Ordinance 173 (1977), and is advocated by Eisenstadt, Long, and Utton in *A Proposed Solar Zoning Ordinance*.

an approach is workable provided the wall height and/or hours are adjusted to local conditions (which may vary from area to area within a city). Although it would be preferable to a pure time option without controls on collector siting (because some incentive to site reasonably is built in), the hypothetical wall option adds nothing but complexity to a space specification of access.

Controlling Collector Locations

If public protection of solar access is extended on the basis of the needs of collectors, without regard to resulting restraints on nearby properties, the restraints on these properties will depend on collector location. Within any particular lot, the higher and farther north a collector is sited, the less it restrains development; the lower and farther south it is, the more it restrains development (see Figure 2-4). For this reason, it would be unwise to protect solar access for all collectors without regard for their siting. Nonetheless, some legislation[21] and proposals[22] have ignored collector location.

A locality could decide to provide solar access in one law and then control collector siting in a separate law, such as the zoning or building code.[23] The problem with this is that if solar access is provided by one law, a second law to control collector locations may not be passed, and some collectors may be located so as to burden neighboring properties unreasonably. (An example would be a collector located on the southern edge of a lot next to an undeveloped lot.) Furthermore, if a separate law were passed allowing collectors only in certain locations, solar users would not be able to make private arrangements with neighbors that would allow them to site their collectors where they would impose greater restrictions on the neighbor's land.

Public protection for solar access could be extended only to collectors reasonably located under the circumstances, and people could be allowed to locate collectors elsewhere if they make private arrangements for more access or if they are willing to take the risk of being shaded. The key problem in this approach is either defining what reasonable locations are or providing a procedure for determining on a case-by-case basis that the location of a proposed

21. E.g., 1977 N.M. Laws ch. 169.

22. Ranging from the federal bills like H.R. 259, 95th Cong., 1st Sess. (1977), to state bills like Colo. S.B. 38 (1976).

23. This is recommended by Sandy F. Kraemer, for example, in SOLAR LAW (Colorado Springs: Shepard's Inc., 1978), pp. 123, 125.

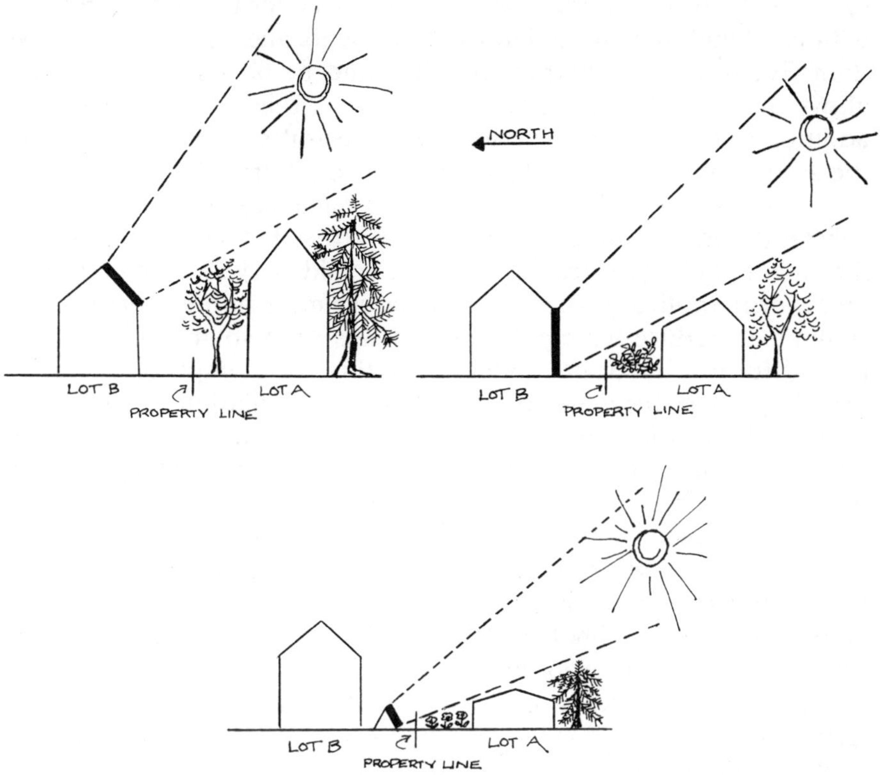

FIGURE 2-4
EFFECT OF COLLECTOR LOCATION ON NEIGHBOR'S DEVELOPMENT
The dotted lines indicate areas needing protection against encroachment. The extent of development allowed on burdened Lot A depends on B's collector placement. Note that Lot A may be much more densely developed when a rooftop collector is used than when the collector is located on the ground.

collector is reasonable under the circumstances. The hypothetical wall and property-line-based plane options contain an inherent incentive for reasonable siting of collectors. If collectors are not reasonably sited, they can be shaded in the former case as much as the hypothetical wall would shade them and in the latter case as long as the development stays under the protection plane.

A simple method of controlling solar loctions has been included in the new California Solar Shade Control Act to protect against shading by vegetation from 10 A.M. to 2 P.M.[24] For a solar collector

24. Cal. A.B. 2321 (passed 31 August 1978).

to enjoy protection under the act, it must be located in compliance with local building and setback requirements and be set back no less than five feet from the property line and be no less than ten feet above the ground. A collector may be lower than ten feet only if, in addition to the five-foot setback, the collector is set back three times the amount it is lowered. (This amounts to a thirty-five foot setback for a ground level collector.)[25] This approach may be reasonable for vegetation control in areas lacking tall trees but would be far too limiting if applied to structure control outside large-lot single-family areas, or if applied to sites where trees are more needed for shade in summer to reduce air conditioning loads than solar collectors are needed for heating or hot water. Since this law exempts only vegetation that shades the collector at the time of installation or during the "remainder of that annual solar cycle,"[26] immature trees planted to minimize future cooling loads might have to be trimmed to protect a neighboring collector.[27] Such trees may not be allowed to grow tall enough to provide beneficial shade. In other words, there are probably climates for which this simple approach to setbacks for protection from vegetation shading would be practical, but for a state as large and varied in climate as California, there are bound to be areas for which this is inappropriate.[28]

Proposed legislation in Massachusetts[29] and the prototype collector recordation ordinance set out in Chapter 7 allow for a case-by-case consideration of whether a proposed collector location is reasonable. Even where a case-by-case approach is to be taken, adoption of regulations defining reasonable location would be helpful both to the administrators ruling on applications and to potential applicants. The prototype envelope ordinance in Chapter 5 uses property-line-based protection planes so that it has inherent control over collector locations.

25. Ibid., § 1 (to be codified as CAL. PUB. RES. CODE § 25982).

26. Ibid., whatever that means as opposed to the following year.

27. There is another unclear section (to be codified as CAL. PUB. RES. CODE § 25986) that exempts a "natural solar . . . cooling system" that shades an active collector; this may possibly exempt some trees, but not if the shadows they cast fall on a passive collector.

28. Fortunately, the act allows localities to adopt an ordinance exempting them from its provisions (to be codified as CAL. PUB. RES. CODE § 25985).

29. Mass. S.B. 304 (1978), Mass. H.B. 5711 (1978), and earlier bills, discussed in Chapter 7.

Evaluating the Options for Specifying Solar Access Protection

Time Versus Planes

For any given collector location it is possible to specify either times of protection or planes of protection that are largely equivalent. But the equivalence is not perfect. When the specification is in terms of time alone, the resulting space that is protected against incursion is defined by a complex surface.[30] This complex surface allows structures and vegetation to be taller due south of the collector site than to the southwest or southeast, because the sun is higher at solar noon than it is earlier or later in the day. Thus, knowing the exact limits on development under a time option requires complicated calculations of both the sun's position relative to the protected collector site(s) and the resulting spaces in the sky that should not be occupied. When there are several collectors nearby, these calculations can require a computer to achieve even moderately accurate results.[31]

The planes of protection described in Appendix A are a crude approximation compared to the complex mathematical surfaces implied in any time option. Those planes do sacrifice some potential for development due south of a single collector in order to use a form of specification that can be more easily understood (especially when in terms of simple rise-to-setback ratios) and enforced. Where there are, or potentially will be, a number of collectors mounted similarly on a row of buildings, the difference under the two methods would be negligible.

Hence, the plane option is more suited to a planning approach that includes protection of future access for a number of sites. The time option *may* be more suited to a case-by-case approach, where few collectors are expected in any particular area. "May" is used here because the complexity of required calculations for each site makes enforcement of the time option difficult before shading actually occurs.[32] It would be more difficult to say in advance whether

30. *See* Appendix C.

31. *See* N. Arumi and R. L. Dodge, *The Formal Implication of Solar Rights,* in ENERGY AND BUILDINGS, vol. 1 (Lausanne: Elsevier Sequoia, 1977), pp. 183–91, for a discussion of large errors that can result from doing the calculations by hand instead of by complex computer programs. Their example uses a fairly complex commercial zone with mixed heights.

32. Of course, once shading occurs, a glance at a watch tells whether it is lawful.

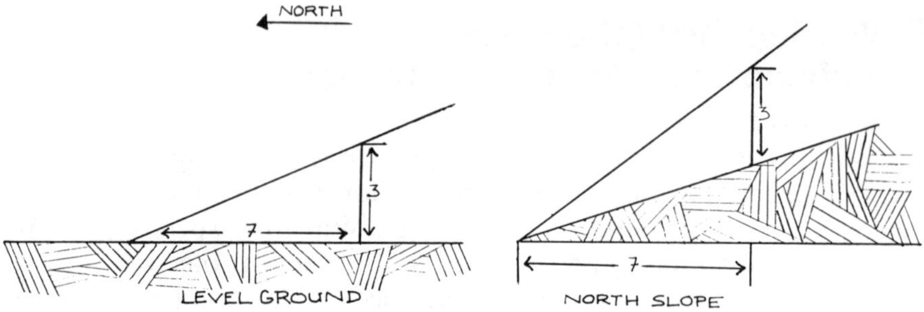

FIGURE 2-5
SOLAR PLANE SLOPE MEASURED TO ACCOUNT FOR NORTH SLOPE
For 3 to 7 plane slope.

a proposed tree or building would unlawfully shade a collector site under the time option than under the plane option.

Another problem with time options, unless the times of protection vary with site slope,[33] is that they invariably are too restrictive of development on north slopes. Shadow lengths are sensitive to site slope and can be much longer on north slopes (the steeper the slope, the longer the shadows). When the restraints on land near protected collector locations are reasonable for level ground under a time option, they are bound to be too restrictive of development on steep north slopes. This situation may be resolved either by graduating the times of protection according to site slope or by using sloping planes that take site slope into account. The former would be awkward, the latter simple, since the plane slope can be measured relative to the slope of the ground (see Figure 2-5).[34]

Hybrid options such as the hypothetical wall have the same general strengths and weaknesses as other time options. The hypothetical wall method does have the advantage of having some built-in control over collector location (collectors too close to the property line may be shaded lawfully).

33. ELI has not seen this proposed, nor is it recommended, though it is possible.

34. Plane slope can be measured relative to the north–south slope of the ground (or east–west slope for planes in that direction), or relative to the horizontal or south slopes to provide for denser development of south slopes while protecting just as much solar access as on level ground. Although this would give poorer protected access on north slopes (access will rarely be good in existing development on north slopes anyway), it seems a reasonable compromise to allow for some further north slope development. The APA GUIDEBOOK FOR PLANNING OFFICIALS recommends access on north slopes equal to that on level sites for new residential developments. ELI does not consider this to be feasible for most existing developments. It is, however, a matter to be considered by a locality in choosing appropriate solar access standards for the various areas within the jurisdiction.

Collector-Based Versus Property-Line-Based Planes

A plane of protection can be specified in relation to either a collector location or a property line.[35] Although the specification looks different, the same plane can be defined from either perspective. If protection from shading is in terms of planes, the use of property lines as the starting point has the advantage in that it does not vary with collector location, and there is no need to control collector locations to avoid unreasonable burdens on neighbors.

If planes are based on collector location, there is need of some control over collector siting to avoid unfair burdens on neighbors. If a collector is sited so high or so far north that a collector-based plane would be less restrictive than a property-line-based plane, it would be possible to use the former instead of the latter where property-line-based planes are normally used. But this would remove the incentive, under a property-line-based scheme, to site as high and as far north as possible to get the most access possible. Under the property-line-based approach, however, the plane heights would have to be worked out in advance for each area to be covered, taking area conditions into account.

Summary of Reasonable Options

For individual private arrangements, whatever is acceptable to the parties in question should be acceptable in the eyes of the law. Under many proposed and adopted solar easement laws, however, only easements stated in terms of angles or planes are recognized.

For a case-by-case approach to public protection of solar access, protection could be in terms of either hours of compensated standard time or collector-based planes. In either case, control over collector siting is essential and is best done in the solar access legislation or regulations promulgated under it to assure coordination.

For a planned approach to public protection of solar access, protection would best be in terms of property-line-based planes, with appropriate heights and slopes chosen to fit each area of the adopting jurisdiction.

Inconsequential Shadows

Very long shadows are cast near dawn and dusk, but little useful energy can be gathered at these times even when a collector is unshaded. To attempt to prevent shadows on collectors from dawn to

35. *See* Figures A-6 through A-11 in Appendix A.

dusk is unreasonable. Unlimited solar access would promote urban sprawl, place unreasonable restrictions on development and might render existing comprehensive plans useless. In spite of this, some proposed and enacted legislation contains no time limitations.[36]

Utility poles, television antennas, wires, and similar obstructions as well as the bare branches of some species of trees cast shadows that have little effect on collector efficiency. Such trifling shadows should be exempted from shading restrictions.[37] Nevertheless, the cumulative impact of many small shadows may reduce collector efficiency. The task of writing and enforcing regulations dealing with minor exceptions is probably best left to the local agency with land use responsibilities.

Other types of harmless shading should also be considered by solar access laws. For example, localities may wish to allow collectors that are not in use to be shaded (e.g., summer shade on space heating collectors). But governments that plan to promote maximum use of solar energy might not want to allow any structures or trees tall enough to shade promising sites for future solar collectors.

36. E.g., 1977 N.M. Laws ch. 169 (effective July 1978); New York S. 8561-A and A. 11384-A (1978).

37. One approach to permissible shading is to disallow shadows that prevent more than a certain percentage of otherwise available energy from reaching a collector. Massachusetts S. 304 (1978) and Oklahoma H.B. 1741 (1978), for example, specify 10 percent. Each of several neighbors could block otherwise available energy. Laws should specify whether this is allowed. If not, a law should state whether the first neighbor to shade may use the entire 10 percent, or whether the right to shade must be shared.

3
Evaluating Solar Access Laws

There are at least nine criteria (discussed below) that can be applied to any type of solar access law or regulation to test its soundness. Laws that meet these criteria are more likely to accomplish their purpose with a minimum of legal controversy and political resistance. These criteria are supplemental to a quality essential to any good law: careful, concise, and unambiguous legislative drafting. While this attribute is extremely important, it is outside the scope of this report.[1]

The criteria raise two important issues: the possible incompatibility of some types of solar access laws with provisions of the U.S. Constitution, and the advisibility of including solar access provisions in a broader general land use plan. A clear understanding of these issues is essential before any solar access protection legislation is proposed.

Attributes of Good Solar Access Laws

Legislators and state and local officials can use the criteria to evaluate solar access legislation and ordinances. It is unlikely that any proposed law or ordinance for protecting solar access can satisfy

1. A helpful guide on legal drafting is REED DICKERSON, THE FUNDAMENTALS OF LEGAL DRAFTING (Boston: Little, Brown and Co., 1965). *See also Improving Government Regulation,* Exec. Order No. 12044, 43 Fed. Reg. 12661 (1978). Section 1 begins, "Regulations shall be as simple and clear as possible . . . . They shall not impose unnecessary burdens on the economy, on individuals, on public or private organizations, or on state and local governments."

the demands of all these criteria. The prototype legislation in Chapters 5 and 7 does not. In general, we found that the "simple," short proposals that we evaluated were unfair to neighbors, inadequate in protecting access, or both; some complexity in drafting appears essential for a good solar access law. Solar access laws that are integrated with comprehensive plans and take varying development situations into account are difficult to design and implement. But simpler, piecemeal solutions cannot be as fair to all affected persons.

The following criteria are not ranked in terms of their importance, because importance will vary with circumstances. The first three points would usually be more difficult to satisfy than the others.

Protection of appropriate amounts of solar access. A law should provide neither a larger nor smaller "place in the sun"[2] than is actually needed. Factors that should be taken into account include topography; height, bulk, and location of structures and vegetation; present and proposed land use controls and comprehensive plans; latitude, climate, and energy use of structures; and orientation of streets and structures. In most urban areas, development rights are very valuable, so that providing more solar access than is needed would encounter justified resistance. The likely future use of solar devices is an important factor in determining what is reasonable access, as are the amount of sunlight that actually exists and the type of solar energy systems that are available, their economics, and their applicability in the region. In effect, this criterion means that solar access laws should be tailored to fit each neighborhood of similarly situated structures.

Clear and fair allocation of costs and benefits. An access law should not result in excessive economic gains or losses to individuals or to local governments.[3] For example, a government should care-

2. The literary expression "a place in the sun," as defined by William Rose Benet, is surprisingly apt to solar access: "a favorable position that allows room for development." THE READER'S ENCYCLOPEDIA (New York: Thomas Y. Crowell Co., 1963), p. 1083.

3. *See generally* DONALD G. HAGMAN AND DEAN MISCZYNSKI, EDS., WINDFALLS FOR WIPEOUTS: LAND VALUE CAPTURE AND COMPENSATION (Chicago: American Society of Planning Officials, 1978). The least expensive alternative is to avoid conflicts altogether. As this goal will not be attainable, ways must be sought to reduce the costs of conflict resolution (see Chapter 13). It has been suggested that when transaction costs are high because of the many people involved, governmental machinery may be more efficient than the private marketplace. *See* R. H. Coase, *The Problem of Social Cost,* JOURNAL OF LAW AND ECONOMICS 3 (October 1960): 1–42. *See also* James Atwood, *An Economic Analysis of Land Use Conflicts,* STANFORD LAW REVIEW 21 (1969): 314–15.

fully consider, and specify in the law, who should bear the costs of vegetation removal or pruning. As with any good law, both direct costs and transaction costs should be kept as low as possible.[4] The benefits and burdens should not be inequitably allocated among classes of landowners and users. Solar access should be available not only to people owning houses on large lots but also to tenants and those living in crowded areas with little or no yard space. Commercial and industrial properties will also require solar access protection.

Compatibility with other policies and laws and with actual conditions in the physical environment. The provisions must be compatible, as far as possible, with existing laws, regulations, comprehensive land use plans, and present land uses. But if the solar access law requires simple additions and deletions to local ordinances or land use laws, this should not be viewed as a serious disadvantage. For example, to create as few nonconforming uses as possible, the provisions of the law should consider the actual conditions within existing developments. Furthermore, the regulations must not be premised on incorrect assumptions, such as that all developments are on level ground, or that all streets are oriented north–south and east–west, or that lots are landscaped only in one particular manner. Nor should the law thwart other important public goals. Access legislation should not unduly reduce the potential tax base provided by real property values. It should promote an overall decrease in the use of nonrenewable energy and should allow dense enough development so that people can live near where they work, study, or shop. Good planning is essential: solar access provision must be integrated with other public goals.

Adequate notice. Clear notice should be given to all affected parties concerning their benefits and burdens. The law's language should be as simple and unambiguous as possible and should not require greater expertise than is needed to understand the descriptions of other regulations on real property. Although the U.S. Constitution requires due notice, solar access is such a new and complex kind of need that it is particularly important to provide more than just constitutionally adequate notice.

4. Transaction costs for private solar access provision could include the expenses connected with bargaining, giving notice, drawing up a contract (lawyers' fees), surveying, and recording fees. For public laws, there would be the cost of land use planning; legal, scientific, and architectural research; costs associated with the legislative process; surveys; mapping; enforcement; etc.

Political acceptability. All of the criteria will contribute to the political acceptability of a legislative proposal. In addition, however, solar access laws must recognize the concerns of local citizens. For example many voters may attach great value to certain types of vegetation, such as particular tree species. To them, the appearance of their neighborhood may outweigh the benefits of greater solar access. Another example where citizen preference may influence a law concerns fencing. In some regions people prefer privacy fences to broad vistas of green lawns. Solar access laws should reflect that preference in their definitions of places to be kept free of shadows. Furthermore, solar access laws should not give people the feeling that their traditional rights are being invaded (requiring permits to plant trees or shrubs, for example, would outrage many people). On the other hand, a proposal that does not go far enough in encouraging the use of solar energy may also encounter political opposition.[5]

Reasonable flexibility. The law should be adaptable to changing societal needs and to changes in solar technology, and should be subject to modification or termination at some future date if this should be in the public interest. For instance, a city may wish to encourage denser development in a single-family-home area close to a central business district or on a new subway route. Solar access laws should not forever freeze the normal development of urban areas.

Another type of flexibility is also important: local governments should be free to decide what specific standards must be met, such as the amount of solar access that should be protected and the kinds of obstructions of solar access that are permissible. It may be desirable, however, for state-level legislation to include guidelines for localities to follow in making these decisions.

Compensation for cancellation of access protection. Along with the need for civic flexibility is a need for individual financial security. If people have relied on solar access laws and purchased solar energy systems, it is only fair that they be compensated if the protection is removed and shading occurs. Without reasonable investment security, many people may be unwilling to invest in solar energy equipment. An alternative would be to allow collector owners enough time to recoup their investment before permitting their previ-

5. Some people feel that they have a "right" to sunshine, even though this position is not supported by present law. For example, the Declaration of Solar Independence states: "[I]n addition to Life, Liberty, and the Pursuit of Happiness, all people are entitled to a place in the sun, and that reasonable solar access may not be denied without due process of law or just compensation." ENVIRONMENTAL ACTION, no. 19 (11 February 1978), 9: 9.

ously protected collectors to be shaded. (This approach is somewhat analogous to allowing nonconforming structures to remain a set number of years.)

Simple implementation and timeliness. Implementation of an access law should not require the establishment of a new state or local agency but should set clear standards for existing agencies to follow in discharging their new responsibilities. Legislation that could be implemented soon is preferable to that which would require long delays. On the other hand, the legislation's effect may be so great that haste must not dominate quality. A community may want to take simple steps immediately (such as facilitating the creation of private easements and redefining the height of buildings) while perfecting a more complex plan.

By clearly defining rights and duties, a solar access law should discourage arguments between neighbors over which party is in compliance. Because some disputes will be unavoidable, the legislation should, where possible, provide for arbitration proceedings or administrative hearings as alternatives to lengthy and costly litigation. (Chapter 13 discusses the arbitration option.)

Protection of future access. As a general rule, ELI favors approaches that protect solar access that will be needed in the forseeable future and that will probably be lost if not secured now. The basic reason for this preference is that it could save money. For example communities with many tall trees could prohibit the planting of any new trees near the northern boundary lines of lots. Such a law would not impose any financial burden now and may avoid heavy expenses for fuel and electricity in the future. Similarly, construction could be permitted only after its impact on future access is considered and steps taken to mitigate any unfavorable impact.

Integrating Solar Access with Land Use Planning

Land use controls shape our cities and many aspects of our lives. They influence how many hours a week we spend in autos, buses, and trains; what time we must rise to travel to work or school; and the distance between our house or factory and our neighbor's. The latter is one of their most visible results. Where lots adjoin and property rights rub against one another, conflicts may arise. Most people are concerned when changes affecting their own property, or that of their immediate neighbor's are proposed. Solar access may affect land use markedly. Unless providing access to sunshine is part of, or at least compatible with, a general land use plan, it may thwart other important societal goals and arouse needless controversy.

The comprehensive plan in general. Comprehensive planning has become a standard practice in urban areas of the United States. The governments of most large cities and many smaller cities, counties, and villages have a planning department. Municipalities that are unable to support planning departments of their own often retain private planning consultants.[6] Plans generally consist of a verbal and graphic description of the community's goals, plans for mobilization of the resources to achieve the goals, and means of implementing the plans through various land use controls like zoning.

Legal requirements for a comprehensive plan. The Standard State Zoning Enabling Act requirement that local governments zone "in accordance with a comprehensive plan" is widely adopted.[7] Until recently, however, courts generally did not require municipalities to adopt a formal, written plan before exercising the zoning power.[8] Courts discovered a "plan" in the zoning ordinance itself, or in the comprehensive extent of zoning regulations.

Within the last decade, a trend has begun toward greater recognition of the plan. Recent legislation has emphasized the planning requirement in the use of zoning and other land use controls.[9] A growing number of courts acknowledge the requirement for a comprehensive plan independent of the zoning ordinance.[10] In certain instances, the lack of a comprehensive plan will weaken the presumption of validity normally granted a legislative action.[11] In addition, a munic-

6. Recent growth of local planning is due in part to encouragement by the federal and state governments. A number of federal and state assistance programs help defray costs of planning activities.

7. U.S. Department of Commerce, Advisory Committee on City Planning and Zoning, A Standard State Zoning Enabling Act (Washington, D.C.: U.S. Government Printing Office, 1926), § 3. The act was drafted and made available during the 1920s under the auspices of the U.S. Department of Commerce, and serves as a model for the zoning enabling statutes of most states, which grant the power to zone to municipalities. It is now in effect (with various modifications) in 47 states; 32 states have adopted verbatim the requirement that zoning be done "in accordance with a comprehensive plan"; nine more states have included analogous provisions in their zoning enabling legislation. Norman Williams, Jr., American Land Planning Law, 5 vols. (Chicago: Callaghan & Co., 1974), 1: § 18.01, p. 355.

8. *See* Kozesnik v. Township of Montgomery, 24 N.J. 154, 131 A.2d 1 (1957); and Udell v. Haas, 21 N.Y.2d 463, 235 N.E.2d 897, 288 N.Y.S.2d 888 (1968).

9. *See* Edward J. Sullivan and Laurence Kressel, *Twenty Years After: Renewed Significance of the Comprehensive Plan Requirement,* Urban Law Annual 9 (1975): 53–63.

10. Ibid., p. 41. *See also* Eves v. Zoning Bd. of Adjustment, 401 Pa. 211, 164 A.2d 7 (1960).

11. *See* Raabe v. City of Walker, 383 Mich. 165, 174 N.W.2d 789 (1970).

ipality that exercises land use controls without adopting a comprehensive plan may be leaving itself open to legal challenges based on spot zoning and lack of equal protection.

The comprehensive plan and solar access. Communities that wish to protect access to sunlight through the exercise of land use controls should adopt a solar plan or plan component as part of their comprehensive plan. Protecting solar access does not call for writing an entirely new plan. But a solar component must be more than an endorsement of the use of solar energy; it must also contain descriptions of the tools that will be used to protect rights to solar access, the policies that will guide the application of these tools to particular parcels of land, and the procedures to be used by those administering the plan.

Adoption of such a plan would not bind the municipality to any future actions. In most jurisdictions a plan is not an instrument of land use control, but a guide to community development that is merely advisory.[12] It will still remain for the legislative body of the municipality to implement the plan through direct controls. The State of Oregon allows county planning commissions to recommend ordinances to protect solar access.[13] There has not been much meaningful activity under this law. Some local governments have gone through the exercise of evaluation, but none has implemented a solar access ordinance. One informed observer blames the delay on a lack of technical assistance and funds.[14]

Basic Legal Issues

Solar access laws must comply with the due process and equal protection clauses of the U.S. Constitution. More specifically, chal-

12. ROBERT M. ANDERSON, AMERICAN LAW OF ZONING, 2d ed., 5 vols. (Rochester, N.Y.: Lawyers Co-operative Publishing Co., 1976), 3: § 21.15, p. 609.

13. Section 2, OR. REV. STAT. 215.110(1)(g). According to Alan D. Kiphut, solar energy specialist for Oregon's Department of Energy, present enabling legislation does not allow *city* planning commissions or city *and* county governing bodies to recommend or adopt solar access ordinances and standards. Senate Bill 299 (1979) would allow local governments to implement these options. *See also* ch. 314, 1978 Conn. Pub. Acts, which encourages "energy efficient patterns of development, the use of solar and other renewable forms of energy" by authorizing local and regional governments to use their planning, subdivision and planned unit development controls, and zoning powers for these purposes. *See further* ch. 786, 1978 Minn. Laws, which enables localities to use zoning to regulate solar access, permits cities to use subdivision regulations for this purpose, and requires a solar access element in all land use plans.

14. Rick Morgan, Oregon Department of Energy, in his review of a draft of this study.

lenges based on the following grounds are likely if access laws are not well-conceived and well-written:

- ☐ Lack of a valid public purpose
- ☐ A taking of property without just compensation
- ☐ Violation of equal protection requirements as reflected in the uniformity clauses in state constitutions, or because of spot zoning

Whether a Public Purpose Exists

Unless courts and legislatures are convinced that developing solar energy is in the public interest, solar access protection through land use controls will not be possible. In other words, there must be a public purpose that justifies any publicly enforced restriction on private rights.

The police power and the eminent domain power are the two main traditional legal fonts of governmental power over privately owned real property. Land use controls must be constitutionally justified on the basis of "public purpose" or "public use." The term "public purpose" is normally used in relation to police power functions such as zoning, while "public use" (or "public benefit") is found in relation to eminent domain. Historically, the public use standard has been the more difficult to sustain. Yet the U.S. Supreme Court's opinion in *Berman v. Parker*[15] suggests that little material difference between the two remains, at least at the federal level. Even assuming that the public purpose test is, in essence, the only standard that a police power control or eminent domain measure must satisfy to justify its use in solar access protection, the burden remains a significant one, worthy of special discussion.

Actions based on the police power. Land use controls will satisfy the public purpose requirement if they promote the public health, safety, morals, or general welfare. The question to be resolved is whether the objective of solar access protection can be sustained in light of these basic police power tests.

Public safety and morals are seldom used to justify land use controls, although safety purposes are an important reason for building codes and many other permit requirements. In any event, safety and morals bear no logical relationship to solar access.

Protection of the public health is the rationale for requiring some types of permits, for preventing the overcrowding of land, and

15. 348 U.S. 26 (1954).

for assuring the availability of adequate light and air. Zoning and use restrictions to achieve these purposes are expressly authorized by the Standard State Zoning Enabling Act and by the zoning enabling acts of nearly all states.[16] The public health purpose is often used to support zoning provisions that also have other purposes. Maximum height regulations, setback requirements, minimum lot size restrictions, yard regulations, and use restrictions, for example, are often justified in part on the basis that they assure adequate light and air.[17]

It is doubtful that a health purpose would, by itself, support a solar access law. (The amount of light present laws provide for health purposes is much less than collectors require to function well.) Although solar energy use may enhance air quality by reducing the need for burning fossil fuels, this benefit will be enjoyed mostly in the air-shed of the power plant, which may be remote from the solar site. A direct benefit to the neighborhood where solar energy was used would provide a stronger rationale. Further, health-related laws usually forbid something that is harmful to health, rather than encourage something that may be beneficial. Access laws and regulations should, nevertheless, list health protection as one of their purposes.[18]

Health and safety protections are relatively specific grants of authority. However, the final type of public purpose—promoting the general welfare—presents some difficulty in interpretation. Since it is often the main justification for the exercise of land use control powers, it needs to be closely examined. Laws and regulations promoting the general welfare of the community are specifically authorized by most zoning enabling acts.[19] Unfortunately, it is not clear how broad a grant of authority this phrase confers, although it is evident that it does not authorize police power regulations for purposes that are only remotely related to the general welfare. Such diverse matters as housing needs, aesthetic regulations, protection of property values, regulation of growth, and the orderly regulation of land all come under the heading of general welfare.

While the general welfare criterion is not all inclusive, it may cover regulatory programs for energy needs. In deciding whether a public purpose exists for solar access laws, courts are likely to con-

16. Anderson, American Law of Zoning, 1: § 7.13, p. 555.

17. Ibid., § 7.05, pp. 544–45.

18. In Japan, the promotion of health is a major reason for guarantees of access to generous amounts of sunlight.

19. Anderson, American Law of Zoning, 1: § 7.13, p. 555.

sider the degree to which the state and local governments have formally committed their police power authority to the goal of encouraging solar energy systems. For example, states should expand the purpose and authority sections of their planning and zoning enabling acts to specify that solar access protection is a public purpose. Solar access protection may survive a legal challenge simply because of the strong presumption of validity that courts since *Village of Euclid v. Ambler Realty Co.* have attached to the purposes of land use regulations claiming to serve the general welfare.[20] It is equally important that local governments revise their comprehensive plans to incorporate solar access protection as a goal. On the other hand, federal legislation stating the importance of solar energy to the nation will probably have little impact on whether state and local solar access laws are upheld. Even state-level policy statements may not greatly affect the validity of a local ordinance.

Eminent domain actions. The use of eminent domain is subject to substantive and procedural tests. The fifth and fourteenth amendments to the U.S. Constitution provide that private property shall not be taken for public use without just compensation and that no state shall deprive any person of life, liberty, or property without due process of law. Most state constitutions contain similar language. Just compensation, public use, and a public necessity for the taking are the three substantive prerequisites to any exercise of this power.

The just compensation requirement will create only administrative difficulties in the implementation of solar access laws. Determining what amount of compensation should be paid in each case will be hard, but no more difficult than in other types of eminent domain cases.

Judicial deference is given to legislative findings that a taking of property is necessary.[21] The reason for deference is the belief that a legislature is best suited to determine the needs of a community and

20. 272 U.S. 365 (1926).

21. *See* Berman v. Parker, 348 U.S. 26 (1954); United States ex rel. Tennessee Valley Auth. v. Welch, 327 U.S. 546 (1946). This deference is generally expressed in a strong presumption that if the legislature has authorized a condemnation, the use to which it is to be put is a public use. *See* United States v. Gettysburg Elec. R.R. Co., 160 U.S. 668 (1896); United States ex rel. Tennessee Valley Auth. v. Welch, 327 U.S. 547 (1946); United States v. Hunting Rights of Swan Lake Hunting Club, 237 F. Supp. 290 (D. Miss. 1964); Martin J. King, *Rex Non Protest Peccare??? The Decline and Fall of the Public Use Limitation of Eminent Domain,* DICKINSON LAW REVIEW 76 (1972): 266, 267.

the means by which those needs can be served. Questions of necessity become a judicial issue only when fraud, bad faith, or abuse of discretion is found.[22] Abuse of discretion is seldom found; fraud or bad faith are also difficult to prove. Courts will generally uphold a taking, therefore, unless the legislature, in authorizing the condemnation, fails to meet the public use requirement. The use of eminent domain to create private legal rights to solar access will be feasible only if courts accept solar access as constituting a truly public use.

The definition of public use has changed over time; different jurisdictions apply different tests. Actual use of the condemned land by the public was commonly required in the middle of the nineteenth century. Most states no longer adhere to this test because of its obvious rigidity, and the actual use test has almost disappeared.[23] To deal more equitably with increasing governmental control over land use, a more elastic concept of public use was required. In *Berman v. Parker* the U.S. Supreme Court said:

> The concept of the public welfare is broad and inclusive . . . The values it represents are spiritual as well as physical, aesthetic as well as monetary. It is within the power of the legislature to determine that the community should be beautiful as well as healthy, spacious as well as clean.[24]

The general rule is that when condemnation is for a private use or to accomplish an end not public in nature, courts will more strictly scrutinize a legislative declaration of public use and may hold it unconstitutional. In some jurisdictions, however, so long as some public advantage or benefit is shown, the power of eminent domain can be used even if private benefits also result.[25] In slum clearance cases, for example, condemnation will be upheld even though the

22. Donald G. Hagman, Urban Planning and Land Development Control Law (St. Paul: West Publishing Co., 1971), § 176, p. 316. *See* United States v. Gettysburg Elec. R.R. Co., 160 U.S. 668 (1896); State ex rel. Sharp v. 0.62033 Acres of Land, 49 Del. 174, 112 A.2d 857 (1955); Kaskel v. Impellitteri, 306 N.Y. 73, 115 N.E.2d 659 (1953) and Wilson v. United States, 350 F.2d 901, 907 (10th Cir. 1965). *Contra,* People ex rel. Dept. of Public Works v. Chevalier, 52 Cal. 2d 299, 340 P.2d 598 (1959).

23. *See* Comment, *The Public Use Limitation on Eminent Domain: An Advance Requiem,* Yale Law Journal 58 (1949): 599.

24. Berman v. Parker, 348 U.S. 26, 33 (1954). *See also* Keystone Collection Service, Inc. v. Recio, 389 F. Supp. 164, 166 (D. Puerto Rico 1975); In re United States, 28 F. Supp. 758, 761 (W.D.N.Y. 1939); Puerto Rico v. Eastern Sugar Assocs., 156 F.2d 316, 323–24 (1st Cir. 1946), *cert. denied,* 392 U.S. 772.

25. Puerto Rico v. Eastern Sugar Assocs., 156 F.2d 316 (1st Cir. 1946), *cert. denied,* 392 U.S. 772; Mt. Vernon Cotton Co. v. Alabama Power Co., 240 U.S. 30 (1916); Clark v. Nash, 198 U.S. 361 (1905); Strickley v. Highland Boy Mining Co., 200 U.S. 527 (1906).

condemned land is later sold to private redevelopers.[26] Nevertheless, courts seem more willing to find too much private benefit in eminent domain cases than in exercises of police power.

For laws protecting access to sunlight, courts will have to be convinced that efforts to reduce the consumption of fossil fuel are essential to community prosperity or will otherwise advance the public welfare. Showing that this requirement is met will be easier if, under its police powers, a city adopts solar access as a general policy in its comprehensive plan and especially if the state adds language to enabling statutes declaring solar access provision to be a proper use of eminent domain powers. Condemnation of individual solar easements that can practically be used only by particular private parties for the benefit of their buildings or homes may still be difficult to justify as a public use. The public benefit seems less concrete than the private. Even cases that adopt the broad public use test concern situations in which the condemnation of property results in a benefit that is enjoyed by a number of people.[27]

It should be remembered, however, that shadows from any particular property can affect other properties to the south, southeast, and southwest in winter, plus the east and west in fall and spring, plus the northeast and northwest in summer. Condemnation of the space over a property may thus affect solar access for several nearby landowners, not just the one due south of the shading property, as is frequently assumed. Relevant points that courts may consider when deliberating solar access situations include the number of property owners using solar devices, the number of structures in a given area that could be retrofitted for solar energy use, and whether the benefits of eminent domain are extended to a large number of landowners or, even better, to all similarly situated landowners.

Possible analogies to the solar access situation are utility line cases in which the public condemns easements across one landowner's property to connect another's building with sewer, water, telephone, or power lines. The analogy of sewer hookups (a health

26. Berman v. Parker, 348 U.S. 26 (1954); State ex rel. Allerton Parking Corp. v. City of Cleveland, 4 Ohio App. 2d 57, 211 N.E.2d 203 (1965), *aff'd,* 6 Ohio St. 2d 165, 216 N.E.2d 876 (1966); New York Housing Auth. v. Muller, 270 N.Y. 333, 342, 1 N.E.2d 153, 156 (1936). Donald G. Hagman notes that courts may show more flexibility as to what constitutes public use in urban renewal cases than in some other situations. URBAN PLANNING AND LAND DEVELOPMENT CONTROL LAW, p. 312.

27. Most prominent have been highway and urban renewal/blight clearance cases.

measure) to solar energy is tenuous, but the analogy to power lines (an energy service) and telephones (a convenience) is somewhat more apt.

Some courts stress the elasticity of the public use concept. They feel that it is no longer a real impediment to the exercise of the eminent domain power.[28] Before such courts, exact analogies may not be needed and solar access may be deemed merely an expansion of what is presently considered a public benefit. However, the incidentalness of private benefits remains as a separate issue in eminent domain cases.

Summary. Regardless of whether "public purpose" or "public use" is the standard against which solar access controls are measured, well-written laws backed by strong legislative findings and integrated with land use plans have a high probability of being upheld.

Whether an Unconstitutional Taking Occurs

The main constitutional limitations on abuse of the police power, as exercised in land use controls, are the fifth and fourteenth amendments' prohibition of governmental takings of property without just compensation. Any land use control that too severely reduces the value of the privately owned land it regulates is in danger of being labeled a taking. If courts find that a taking has occurred, they may order the government to compensate the owner of the restricted land. A poorly conceived solar access law could so greatly restrict the development of a particular parcel that a taking would happen. (Good drafting can easily avoid this consequence.) State constitutions have their own prohibitions against takings, which may be interpreted somewhat differently from the federal law. The following discussion deals only with federal limitations.

Exactly what courts will call a taking is difficult to predict and will vary greatly with the circumstances. Taking objections are less persuasive if courts believe that an important public purpose is being

28. Johnson City v. Clominger, 213 Tenn. 71, 73, 372 S.W.2d 281, 284 (1963); Comment, *The Public Use Limitation on Eminent Domain*, p. 608. "Many courts have recognized the folly of attempting to lay down any hard and fast rule [as to what public use means] and, while repudiating the dangerous doctrine that any enterprise which indirectly promotes the public welfare is necessarily a public use, have not attempted to confine the exercise of eminent domain to cases in which the public will have the right to use the property sought to be taken." JULIUS L. SACKMAN, NICHOLS' THE LAW OF EMINENT DOMAIN, rev. 3d ed., 14 vols. (New York: Matthew Bender, 1976) 2A: § 7.2[3].

served by the regulatory program.[29] In a July 1978 U.S. Supreme Court ruling, *Penn Central Transportation Co. v. New York City,*[30] which concerned development rights above Grand Central Station, the Court said that historic preservation measures are analogous to zoning actions and should be evaluated under the same legal tests. The Court insisted upon looking at the effect of the regulation on the property as a whole, refusing to allow the taking analysis to focus only on the air rights. This aspect of the decision is particularly relevant to solar access because the key question is whether any beneficial use of the parcel as a whole remains: the Court distinguished the *Penn Central* case from *U.S. v. Causby*[31] on the basis that *Causby* involved an invasion of airspace (by aircraft) that destroyed the use of the farm beneath. In *Penn Central,* New York City did not impair the present use of the landmark railway terminal. Nor did the government appropriate any property for its own use. Regulations providing solar access are more analogous to *Penn Central* than to *Causby*.

It is difficult, however, to envision a solar access law that would regulate just a single property, as was the case in *Penn Central.* Where landmark preservation laws apply, at most, to a handful of structures, solar access laws could eventually protect and restrict thousands. The extent to which all landowners in an area share the benefits and burdens of the regulatory action is an important consideration. At the nub of the landmarks problem is that the individual landmark owner is singled out to bear an often drastic burden

29. The proposition is illustrated by Modjestka Sign Studios, Inc. v. Berle, 43 N.Y.2d 468, 402 N.Y.S.2d 359, 373 N.E.2d 255 (1977), where a billboard ban was attacked as a taking by a person who had hoped to lease his land for billboard purposes. The New York Court of Appeals said that zoning for aesthetic ends is a proper public purpose. But the court deemed aesthetic purposes to be less important than public safety, and therefore found a claim for compensation to be more credible (if safety had been involved, such a claim would be frivolous). Case remanded for further consideration of whether the amortization provisions of the ordinance were reasonable.

30. *See* 46 U.S.L.W. 4856, 8 ENVIR. L. REP. 20528 (1978) for the text of the decision and ibid., pp. 10149–53 for an analysis.

31. 328 U.S. 256 (1946). The case concerned a poultry farm adjacent to an airport used by military aircraft. The landing flight path was directly over the farmhouse and chicken house. The people couldn't sleep and the chickens were flying into walls in fright and dying. The occupants finally had to give up chicken farming. The Court held that a landowner owns at least as much of the airspace above the ground as he can occupy or use in connection with the land. Ibid., p. 261. Flights over private land are not generally a taking, however, unless they are so low and so frequent as to be a direct and immediate interference with the enjoyment of the land. Ibid., pp. 265–66.

while his neighbors and the public generally reap the benefit of his loss. There is little or no play, therefore, for what the courts have labeled the average reciprocity of benefit (and burden) principle.[32] Traditional zoning typically avoids this problem because, theoretically at least, all the landowners are equally benefitted and burdened. For instance, while one landowner in a single-family residential zone may not build a hotel, neither may any other.

Rezoning for solar access seems more akin to conventional zoning than to landmark designation, in that all landowners within the zone (except those already in deep shade) are potential users of solar collectors and hence can benefit from the very restrictions that they may claim are offensive. But it would be helpful to know what the likelihood is that these potential beneficiaries will in fact benefit. Under conventional zoning, it can hardly be argued that owners of homes in single-family zones are not in fact benefitted by the stability of land use patterns introduced by the zoning. But suppose that only 15 out of 200 landowners actually install solar energy systems in an area downzoned to protect solar access. Will the fact that the other 185 could have availed themselves of the restriction's benefits convince a court that the restriction honors the reciprocity principle?

While it is true that zoning measures will not be invalidated solely because they burden some landowners more than others,[33] it is difficult to predict where courts will draw the line between the valid and invalid. The most questionable situation would occur under a solar access law that was not related to zoning or planning and that created a right to sunshine whenever and wherever a collector was installed. Imagine that two contiguous vacant lots owned by A and B remain in a high-rise commercial area. Lot owner A, to the north, installs a ground-level collector on his southern lot line. B is effectively precluded from any reasonable development on his lot. The impact on B would be so severe and so unique that a court would most likely declare the law void as beyond the police power, or require government compensation to B.

Another distinction between the factual situation in *Penn Central* and solar access cases is that it is doubtful that courts would

32. For an excellent analysis of the principle and its nonapplication in the landmark context, *see* Penn Central Transp. Co. v. City of New York, 42 N.Y.2d 324, 366 N.E.2d 1271, 397 N.Y.S.2d 914 (1977) (Breitel, J.).

33. Penn Central Transp. Co. v. City of New York, 46 U.S.L.W. 4856, 8 Envir. L. Rep. 20528 (1978).

view the landowners whose development rights are cut down by a reasonable solar access law (as opposed to the example just discussed) as being as grievously disadvantaged as the owner of a landmark building. Assume, for instance, that development is limited to two stories instead of five in a certain neighborhood to protect solar access. Property owners lose their former zoning entitlement to protect their neighbors' solar access. As has been clear at least since *Village of Euclid*, however, mere loss of a profit expectancy does not establish a legal claim to compensation. Downzoning from five to two stories in an apartment-residential zone would not of itself appear to overreach the police power because the affected properties would continue to enjoy a reasonable beneficial use. The U.S. Supreme Court explicitly stated in *Penn Central* that a partial loss in value, standing alone, cannot establish a taking, at least as to land use regulations reasonably related to promotion of the general welfare. A taking occurs only if an owner is deprived of all reasonably beneficial use of the property.[34]

To summarize, in deciding whether a taking has occurred courts look at the severity of the loss in value,[35] the extent to which benefits and burden are shared by all landowners, and whether the restriction provides a real contribution to public welfare. Some courts apply a balancing test to these considerations.[36] A law is considerably buttressed from attack if it promotes goals contained in a comprehensive plan.

Legal Problems in the Absence of a Comprehensive Plan

Solar access laws that are not part of a comprehensive plan are not only more vulnerable to charges of a lack of public purpose, but may also be susceptible to lawsuits alleging a "spot zoning" or a violation of the uniformity requirement in a state zoning enabling act. "Spot zoning" is a judge-made doctrine intended to give courts a way to invalidate certain types of zoning changes described below.

34. 46 U.S.L.W. 4856, 8 Envir. L. Rep. 20528 (1978). But the *Penn Central* decision did not go so far, however, as to hold that any regulation is valid as long as the government does not actually take title to the property or actually enter it.

35. Penn Central Transp. Co. v. City of New York, 46 U.S.L.W. 4856, 8 Envir. L. Rep. 20528 (1978); Cherry Hills Village v. Trans-Robles Corp, 181 Colo. 356, 509 P.2d (1973); Corthouts v. Newington, 140 Conn. 284, 99 A.2d 112 (1953); Arverne Bay Constr. Co. v. Thatcher, 178 N.Y. 222, 15 N.E.2d 587 (1938).

36. Schwartz v. Lee, 50 Misc. 2d 533, 270 N.Y.S.2d 855 (1966); Grand Trunk Western R.R. Co. v. Detroit, 326 Mich. 387, 40 N.W.2d 195 (1949); La Salle Nat'l Bank v. Chicago, 5 Ill. 2d 344, 125 N.E.2d 609 (1955).

Uniformity requirements, on the other hand, are imposed by statute. (Lack of uniformity with adjacent land is one test courts use to determine if spot zoning has occurred.) Both objections are rooted in the equal protection clause as violations of the general principle that properties that are similarly situated should be treated alike.

Spot zoning. Solar access laws are unlikely to be vulnerable to charges of illegal spot zoning if they are part of a general development plan. Spot zoning charges can be used to contest zoning regulations or zoning maps. To identify spot zoning, courts look for one or more of the following:[37]

- □ One or a very few lots are rezoned
- □ The newly allowed use is markedly different from (and perhaps even incompatible with) surrounding uses
- □ The private benefit from the change is greater than the public benefit

Spots are usually rezoned to a less restrictive use, but some cases involve changes to more restrictive classifications.

The question of special benefit is often the controlling issue; if a special benefit is improperly given to an individual to the detriment of the public interest, the other conditions are not needed for a finding of spot zoning. Solar access laws must, then, benefit the community interest and avoid conferring improper individual benefits.[38] The number of people in a given area who benefit, the uniformity of the benefits, and the severity of the restrictions imposed would be relevant factors.

In cities where great care is taken to integrate solar access laws with a general plan and to give all similarly situated lots equal amounts of solar access protection, spot zoning is not a valid charge.

37. Williams, American Land Planning Law, 1: § 27.02, p. 563. Typically, the spot involved is allowed a more intensive use than is generally permitted in the area. Solar access protection, on the other hand, would put more restrictions on the use of properties adjacent to collectors. These restrictions may or may not result in less intensive use of land.

38. An example of how spot zoning could occur in solar access legislation is if a city amended its zoning laws to create a special solar access zone that encompassed only one building. Or suppose a city has a comprehensive plan that calls for very dense development of forty stories for areas around the city hall. In spite of this plan, the city council enacts a new solar overlay zone that protects solar access only for a short row of expensive townhouses on Contributors' Street, which runs behind the city hall. The new law means that the lots for several blocks behind the townhouses cannot be developed much above their present two stories. Their value plummets. Spot zoning *may* have occurred.

One legal scholar has said that "spot zoning" and "zoning in accordance with a comprehensive plan" are "two sides of the same coin. That is, if a given zoning change is in accordance with a comprehensive plan, then it is not spot zoning, and if it is spot zoning, then it is not in accordance with a comprehensive plan."[39] Most courts, when examining a zoning amendment that has been challenged as spot zoning will uphold the amendment if it serves the general needs or policies of the community, and if the proposed change will benefit others in addition to the landowner receiving the amendment.[40]

Uniformity requirements. Another easily avoided difficulty is presented by the uniformity requirements of state zoning enabling statutes. The Standard State Zoning Enabling Act requires the same zoning rules to apply to all properties within each class of uses and within each district. Evidence of dissimilar factual circumstances must justify dissimilar treatment. The uniformity issue may arise in the rare situation when a zoning restriction is applied in only a specified part of a district. Courts usually have found such arrangements invalid.[41] If the land involved differs in character because of the potential for solar energy use or another distinguishing characteristic, however, the problem may often be avoided by dividing the district into two separate zoning districts or by creating distinct overlay zones.

In some states, the uniformity requirement operates much as the comprehensive plan doctrine to ensure consistent policy in mapping. And courts have often used the uniformity requirement as a reason to strike down permits granted without adequate standards or guidelines.[42] For instance, if a municipality passes an ordinance that allows the zoning board to provide solar access protection on a lot-by-lot basis, with no overall strategy or guidelines, the municipality's actions may be challenged as a violation of uniformity requirements (and also as an invalid delegation of power).

When examining a zoning ordinance that has been challenged as lacking uniformity, courts are often unwilling to examine the poli-

39. WILLIAMS, AMERICAN LAND PLANNING LAW, 1: § 27.01, p. 561. *See also* Penn Central Transp. Co. v. City of New York, 46 U.S.L.W. 4856, 4864, 8 ENVIR. L. REP. 20528, 20535 (1978).

40. Daniel R. Mandelker, *The Role of the Local Comprehensive Plan in Land Use Regulation,* MICHIGAN LAW REVIEW 74 (1975): 932.

41. WILLIAMS, AMERICAN LAND PLANNING LAW, 1: § 31.04, pp. 611–12.

42. *See* WILLIAMS, AMERICAN LAND PLANNING LAW, 1: § 31.06, p. 615.

cies underlying a comprehensive plan. Many courts also defer to plan policies when reviewing a particular zoning ordinance.[43] By providing an overall public strategy, a plan provides an important defense against equal protection types of challenges to solar access protection strategies.

43. Mandelker, *Role of the Local Comprehensive Plan*, pp. 926–27.

Part II
Area-Wide Protection

The solar access strategies in this section share three qualities:

- □ They are applied on an area-wide, rather than a lot-by-lot, basis.
- □ They protect access in advance of actual individual need.
- □ They rely on land use controls that are applied to all similarly situated parcels, whether or not collectors are in use on each parcel.

In addition, most of these strategies almost force thoughtful land use planning before they can be implemented.[1]

Area-wide protection is a good solar access approach where it is possible to identify and make generalizations about likely locations for collectors in particular neighborhoods. If generalization is not possible—if the geography, vegetation and structures vary widely—the advantages of areawide protection are less compelling.

Zoning is frequently suggested as a device for providing solar access. Familiar, traditional zoning tools can be modified to provide more solar access protection. A less widely used device is envelope zoning—a device that is well-suited to solar access protection.

There are many reasons why zoning-based approaches are attractive as solar access controls:

1. *See* Chapter 3.

- ☐ Zoning has a long history of use and judicial acceptance. The right of governments to impose zoning controls has been widely upheld by courts for more than fifty years.[2]
- ☐ Zoning already regulates many particulars of property development that are key to solar access: the maximum height of structures; the distance that they must be set back from the front, side, and rear boundaries of their lots; their orientation; what accessory structures are allowed and where they may be put; and the permitted uses in each area.
- ☐ Zoning is often tied to a comprehensive development plan.
- ☐ Of all possible methods of protecting solar access, it is possible that using zoning could place the least demand on the individual solar user. For instance, he or she would be spared the effort and expense involved in applying for permits or purchasing private easements.
- ☐ Developed areas undergo slow but significant changes in character. Zoning ordinances can have more influence on these changes than any other laws and can allow for rational planning, changes in plans, and responses to changes in conditions more readily than approaches that give protection on a lot-by-lot basis.
- ☐ Zoning for solar access could almost certainly comply with the due process and equal protection requirements of the U.S. Constitution and state constitutions.[3]
- ☐ Once in place, most zoning for solar access protection would cost little to administer even when solar installations become numerous.

2. Since Village of Euclid v. Ambler Realty Co., 272 U.S. 365 (1926). The vast majority of cities with populations over 5,000 have established zoning machinery and are accustomed to this type of land use control. In 1960, over 90 percent of cities with populations of 5,000 or over had zoning. NATIONAL COMMISSION ON URBAN PROBLEMS, BUILDING THE AMERICAN CITY (Washington, D.C.: U.S. Government Printing Office, 1969), pp. 208–09. Occasionally, zoning is implemented at the state level, however.

3. As construed, the due process clause requires an ordinance to be reasonably related to a permissible state public purpose. *See* purpose clauses of prototype legislation and accompanying commentary in Chapter 5, and discussions of public purpose and equal protection in Chapter 3.

- Zoning can protect solar access for future installations, instead of allowing development to shade more and more prime sites for future collectors.

Zoning should not be regarded as a solar access panacea in developed areas, however. It has some noteworthy limitations in this regard:

- Like other approaches to solar access, zoning must satisfy the public purpose test, and it is not clear that it will do so in all situations.
- Solar access amendments to state zoning enabling legislation would be advisable, particularly if vegetation is to be controlled. Amendment of the enabling statutes may not guarantee that a court would find a legitimate public purpose, but it would be persuasive.
- In most parts of the nation, zoning ordinances are easily and quickly changed,[4] and some commentators feel collector owners would not be willing to rely solely on zoning for protection.
- Some of zoning's flexible new techniques, such as planned unit and mixed use developments, may be useful in new areas or large-scale redevelopments. However, basic changes in existing developmental patterns—in the types and placement of structures and vegetation—can only be made very slowly, if at all.
- If an access law made many trees or buildings nonconforming uses, it would encounter political opposition and generate a costly administrative workload.

4. This is not the case in some New England areas, which require two-thirds of those at a town meeting to approve a zoning change.

4
Traditional Zoning

Traditional zoning tools and procedures that can be modified to protect solar access include low density zoning; new height, grade, and setback rules; floor-to-area ratio (FAR) controls; downzoning; variances; and overlays.

Under the most familiar type of zoning, each district is controlled by a fixed set of use, height, setback, density, and building coverage regulations. It is not recommended that localities rely heavily on such rules for solar access protection in developed areas because the same numbers (for heights, setbacks, etc.) are applied uniformly throughout each district so that little allowance is ordinarily made for unique site characteristics or for the relationship of a given parcel to nearby development. For example, the direction in which land slopes is critically important to solar access. A building on a hilltop or one on a site that slopes down to the south often enjoys good solar access regardless of activities on adjacent land. But a structure on a lot sloping to the north may be shaded unless severe development restrictions are put on land to its south. (It will be undesirable to try to protect solar access in some of these cases.) However, since many communities developed around, and continue to rely upon, traditional zoning districts, they should seek ways to adapt them to this new purpose. The strategies in this chapter offer an improvement over present laws, but they are best used only as a supplement to other approaches.

The strategies will require the amendment of zoning regulations. Some analysts have said that calling for changes in zoning is politically unacceptable, but there is evidence to the contrary. Municipalities frequently make substantial revisions in their zoning laws. Of cities with over fifty thousand inhabitants, more than half either enacted or made major revisions in zoning laws between 1960 and 1967.[1] This suggests that voters do accept zoning changes. It may be that if citizens perceive the changes to be in their own interest, they will accept them even more readily. High electricity, oil, and gas bills may build a constituency for solar access regulations, even if the regulations would place some further limits on development.

Amendments to zoning laws are typically made by the same legislative body that is delegated the power to adopt the original zoning ordinance.[2] Courts do not have the power to amend zoning laws. It is not necessary to rezone an entire city or county—piecemeal changes are often valid.[3]

All rezoning actions must comply with statutory and constitutional procedural requirements. These requirements—stringently interpreted—are adequate notice (publication in a local newspaper, posting on property, and possibly notice to individual landowners) and an open hearing conducted so that parties affected by the proposed zoning change have an appropriate opportunity to be heard and to present and rebut evidence before an impartial tribunal. A record of the hearing should be made and adequate findings executed.

If an amendment has been properly enacted courts will accord it the same presumption of validity that is given to the initial enactment of a zoning ordinance. A property owner has no vested right in a zoning classification,[4] but does have the right to rely on a rule that

1. ALLEN D. MANVEL, LOCAL LAND AND BUILDING REGULATION: HOW MANY AGENCIES? WHAT PRACTICES? HOW MUCH PERSONNEL? National Commission on Urban Problems Research Report no. 6 (Washington, D.C.: U.S. Government Printing Office, 1968), p. 31.

2. For a discussion of a legislative body's powers to amend and limitations on that power, *see* ROBERT M. ANDERSON, AMERICAN LAW OF ZONING, 2d ed., 5 vols. (Rochester, N.Y.: Lawyers Co-operative Publishing Co., 1977), 1: §§ 4.26, 4.27, pp. 233–41.

3. *See* notes 18 and 19.

4. Conrad N. Bagne, *Up and Down the Zoning Scale (Part I);* LAND USE LAW AND ZONING DIGEST 26, no. 10 (1974), p. 8. *See* HFT Ltd. v. Superior Court of Los Angeles County, 15 Cal. 3d 508, 125 Cal. Rptr. 365, 542 P.2d 237 (1975). The California Supreme Court held that an inverse condemnation action would not lie in a situation where it was alleged that a zoning change resulted in a substantial decrease in the value of a tract of land. (An inverse condemnation suit is an action seeking a condemnation award on the ground that a governmental restriction has effectively taken away the use of the land and is in reality a condemnation.) *See also* ANDERSON, AMERICAN LAW OF ZONING, § 4.26, pp. 238–41.

his zoning status and that of his neighbors will not be changed unless required for the public good.[5] In other words, there must a public purpose to justify rezoning. Since courts will generally not inquire into the motives of board members,[6] the task of proving that an action was arbitrary or capricious is extremely difficult.

Low Density Zoning

In new developments or large scale redevelopment, the simplest way to protect solar access is to require such large lots (in relation to the structures permitted on them) that structures cannot cast shadows across lot boundaries during hours when solar energy is to be protected. This easily administered approach could be accomplished under traditional Euclidian zoning. Of course, new regulations controlling vegetation heights would often be necessary. Large lot requirements, however, promote urban sprawl and are increasingly vulnerable to charges of exclusionary zoning.

Redefining Height, Grade, and Setback

Height. Height limits will be an important element of any solar access law. Most existing height regulations, however, will not be adequate. In areas zoned for single family residences, for example, a uniform thirty-five-foot height limit is fairly common. One analyst notes that "The height limit is the same on a 50-foot lot with 5-foot side yards as on two-acre lots with 25-foot side yards, apparently on the theory that the well-to-do require more light and air than the rest of the population."[7] He concludes that height limits for multifamily buildings are equally arbitrary and bear almost no rational relationship to the principal purposes of height controls: protection from fire hazards, protection against aviation accidents, protection of view, protection of the character of the neighborhood, and protection of light and air.[8]

In spite of their apparent irrationality, height limits would appear to afford a considerable measure of solar access protection against shadows cast by structures. Actually, both the usual defini-

5. *See* Norbeck Village Joint Venture v. Montgomery County Council, 254 Md. 59, 254 A.2d 700 (1969).

6. Anderson, American Law of Zoning, § 4.18, pp. 217–21.

7. Frederick H. Blair, Jr., Height Regulation in Residential Districts, Planning Advisory Service Report no. 237 (Chicago: American Society of Planning Officials, August 1968), p. 2.

8. Ibid., pp. 2–4.

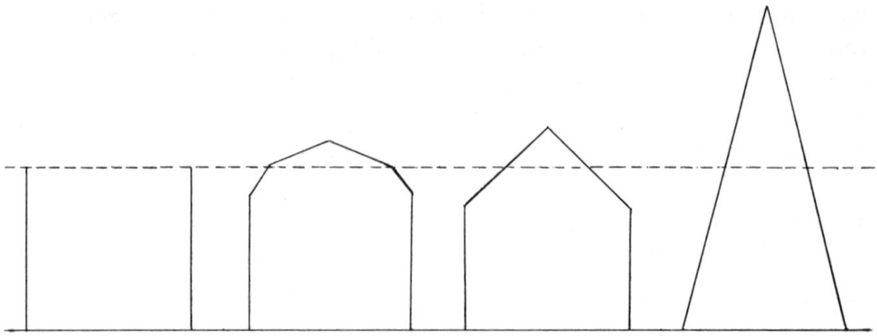

FIGURE 4-1
LEGAL "HEIGHTS" OF BUILDINGS
All of these buildings have the same height under the usual zoning definitions.

tion of building height and various height exceptions cut away much of the protection.

For example, most zoning ordinances define height of a building more or less in the following terms:

> Height of building: the vertical distance from the grade at the front of a building to the highest point of the coping of a flat roof between the eaves and a ridge or to the deckline of a mansard roof or the average height between the plate and ridge of a gable, hip, or gambrel roof.[9]

Figure 4-1 illustrates the meaning of this definition: If a building has a sloped roof, it is taller than its legally defined "height." In turn, the shadows cast by a building of "maximum legal height" would depend on the shape and slope of its roof. Thus, a seventy-foot-tall A-frame building is only thirty-five feet in "height" under this definition, but it would still cast the shadow of the seventy-foot building that it is.

This kind of height definition is clearly unsatisfactory from a solar access viewpoint.[10] A good minimum step for a locality to take toward assuring solar access protection would be to redefine height of structures as the distance from grade to the highest part of the structure, whether that be the roof or something mounted on it, except as expressly otherwise provided. (State enabling legislation would not be required.)

9. Albuquerque, N.M., Zoning Regulations § 4 (1974).

10. Many zoning ordinances set a limit on the number of stories a structure may have. Story regulation poses the question of what constitutes a whole story and what constitutes a half story. For solar access purposes, regulating the number of stories seems unnecessary if height regulations are drawn properly.

Another problem in height regulations is that there are invariably sizeable exceptions. A typical ordinance might provide:

> The building height limits set forth in this chapter shall not apply to belfries, chimneys, cupolas, domes, flagpoles, flues, monuments, radio towers, television antennae or aerials, spires, tanks, water towers, water tanks, air conditioning units or similar roof structures and mechanical appurtenances, except where such structures are located within an airport approach area, as designated on the zoning map. No such roof structure, however, shall have a total area greater than twenty-five percent of the roof area; nor shall such structure be used for any purpose other than a use incidental to the main use of the building.[11]

Exceptions vary from locality to locality. For example, the maximum roof-structure coverage can be larger or smaller than the 25 percent in the above example. In some places it is a stated percentage of the lot size.[12] Some excepted items, such as flagpoles and radio towers, have an unobjectionable impact on solar access. (Incidentally, solar collectors should be in the class of general exceptions or the law may unintentionally prohibit the retrofitting of roof collectors.) Other roof structures, however, could cast objectionable shadows. For instance, a 10-foot-high roof structure with its long side along the north edge of a flat roof could cast a shadow as large as another story, while a structure of the same volume arranged relatively compactly near the southern edge of the roof would cast only a small shadow, perhaps limited to the building's own roof. Therefore, zoning regulations intended to protect solar access must take into account not only the shape and size, but the location of large-scale roof structures, including solar collectors. From a solar access viewpoint, the most reasonable way of defining total exceptions from the height limit might be in terms of some percentage of lot width. The part of the structure requiring exception should generally have to be located as far south and as centered on the east–west axis) as possible, since this relates to how much it would shade neighboring property.

Grade. The definition of grade is also important to solar access protection. In the sample ordinance above, the height is relative to the grade along the front or street side. In some ordinances the grade is the average undisturbed level of the ground on all sides of the building. On level ground, it makes no difference which of those definitions is used. On sloped ground, it matters greatly: If the lot

11. MONTGOMERY COUNTY, MD., ZONING ORDINANCE § 59–18 (1975).

12. For example, 15 percent of lot area in Loudoun County, Virginia. LOUDOUN COUNTY, VA., CODE § 520.4.2 (1972).

slopes down away from the street, a taller building is permitted by the street level reference. On lots that slope down toward the north, a street-side reference point may mean that more structures are severely shaded than necessary. Communities should examine, and if necessary revise, their definition of grade to help achieve the desired level of solar access protection.

Setbacks. A community should also examine and consider changing its building setback requirements from front, back, and side lot lines. For solar access purposes, it makes sense to allow structures (or vegetation) to be sited more flexibly than allowed under most zoning regulations. However, aesthetic or safety concerns may sometimes argue against this. In areas that are already developed, the few new structures that are built may appear out of character with the neighborhood if their placement is very different from that of surrounding structures. The solar access benefits of new setbacks may be minimal unless there is substantial redevelopment or expansion of existing structures.

FAR controls. Rather than using height and setback restrictions, some jurisdictions have floor-to-area ratio (FAR) regulations to control the size of structures. FAR restrictions are particularly common in commercial areas. A FAR of 1.0 (100 percent) means that if a building is only one story tall, it may cover an entire lot. A two-story building, on the other hand, could only cover one-half the lot.[13]

FARs have both advantages and disadvantages with regard to solar access controls. Their disadvantage is that the height and exact siting of a structure cannot be predicted, making it difficult for neighbors to plan their solar energy systems. This same flexibility of FARs could also allow structures to be oriented, located, and shaped in ways consistent with good solar access. Under present FAR regulations there is no assurance that solar access needs will be considered, however.

It may be possible to modify FAR controls to encourage or mandate the accommodation of access needs, by allowing density bonuses in exchange for maximizing solar access. Density bonuses may be optional, as is common, or mandatory.[14] Basically, density

13. Similarly, a FAR of 0.25 would allow a one-story building to cover only one-fourth of a lot, and a two-story building could cover one-eighth.

14. *See* Corbin Crews Harwood, Using Land to Save Energy (Cambridge, Mass.: Ballinger Publishing Co., 1977), pp. 118–23, for a discussion of legal issues related to the use of density bonuses.

bonuses trade additional density for something else desirable to the community, such as public transportation stops, low-income housing, or recreational land. The provision of solar access could be included as an acceptable trade. Higher density does not automatically mean reduced solar access—careful design and placement of structures can greatly mitigate the impact of shadows.

Rezoning to a Lower Density

Rezoning to a lower density, or "downzoning," occurs when a local government amends its zoning ordinance to reduce the potential level of development on particular land parcels.[15] The amendment replaces the parcels' original zoning classification with a more restrictive designation. Lower height limits may be set, or larger setbacks or more restrictive FARs. (No examples could be found in the legal literature, however, where the term downzoning was applied to vegetation controls.)

Downzoning differs from other forms of rezoning in that it is generally initiated by a local government rather than by the owner of an affected property. A municipality may, for example, seek to maintain control over development by rezoning an area at less than its developable capacity. This creates a bargaining position for the community as development rights may be traded for various amenities.[16] Or downzoning may be used to control a development problem that has arisen since an earlier zoning designation was established.[17]

To protect existing solar access in developed neighborhoods, downzoning may be used to establish more stringent restrictions on building height, setback lines, and density. Assume that existing zoning in an area allows three-story structures but that nearly all structures built to date have been one story, and none has exceeded two stories. The extra allowed story or two may block the solar access of many existing, planned, or potential solar collector locations. A downzoning action could be used to lower the maximum allowable height to one story (or two stories). It is not recommended, however,

15. *See generally* Bagne, *Up and Down the Zoning Scale (Part I),* idem, *Up and Down the Zoning Scale (Part II),* LAND USE LAW AND ZONING DIGEST 26, no. 11 (1974); R. Robert Linowes and John J. Delaney, *Downzoning: And How the Landowner May Fight It,* REAL ESTATE LAW JOURNAL 5 (1976): 311–37.

16. *See* Daniel R. Mandelker, *Downzoning to Control Growth Draws a Close Look by the Courts,* REAL ESTATE LAW JOURNAL 3 (1975): 402. *See also* the discussion of transferable development rights in Chapter 11.

17. Bagne, *Up and Down the Zoning Scale (Part I),* p. 6.

that areas be downzoned so severely that many nonconforming structures result. Such an action would be politically unpopular and create a great administrative burden.

Legal issues. Downzoning is somewhat more susceptible to legal attacks if it is applied only to a small area in a "piecemeal" manner.[18] If a community downzones a large area of land based on a comprehensive reconsideration of the factors involved in zoning, state courts will ordinarily apply only a limited judicial review.[19] A piecemeal downzoning is more vulnerable to invalidation on procedural grounds,[20] and also on grounds that it is arbitrary and capricious, that it bears no relation to the public health, safety, and general welfare, or that it is an unconstitutional taking of property with-

18. For the purposes of this study, "piecemeal rezoning" is defined as an action to change the zoning classification of a small area of land without comprehensively rezoning the entire community based on reconsideration of all factors that influenced the original zoning scheme. The same scope of review generally applies to downzoning and upzoning actions. Problems may arise in Maryland and Virginia over so-called "change or mistake" rules. Board of Supervisors v. Snell Constr. Corp., 214 Va. 655, 202 S.E.2d 889 (1974); Board of County Comm'rs v. Edmonds, 240 Md. 680, 215 A.2d 209 (1965). The Virginia rule differs from the Maryland rule in two important respects: (1) the Virginia rule is limited in application to piecemeal downzoning, while the Maryland rule applies to all piecemeal rezoning; and (2) the Virginia rule does not require the governing body to produce *strong* evidence, as Maryland does, but evidence sufficient to render the issue of reasonableness "fairly debatable." *See* Board of Supervisors v. Snell Constr. Corp., 214 Va. 655, 659, 202 S.E.2d 889, 893 (1974). This rule requires that there be evidence of a mistake in the original zoning or a change in conditions to sustain a piecemeal change of the zoning ordinance. Ibid. It is not clear whether a court might define "change of conditions" flexibly enough to include not only physical changes, but also changes in the availability of energy sources and advances in technology.

19. *See* Norbeck Village Joint Venture v. Montgomery County Council, 254 Md. 59, 254 A.2d 700 (1969); and Daniel R. Mandelker, *The Role of the Local Comprehensive Plan in Land Use Regulation,* Michigan Law Review 74 (1976): 922–24. Even piecemeal rezoning decisions are subjected to a very limited review. Michael S. Holman, *Zoning Amendments: The Product of Judicial or Quasi-Judicial Action,* Ohio State Law Journal 33 (1972): 140, 131–32.

20. Although a minority position, an important state court decision provides an example of the type of problem that may arise if a community attempts to rezone on a piecemeal basis. The Supreme Court of Oregon in Fasano v. Board of County Comm'rs, 264 Or. 574, 507 P.2d 23 (1973), suggested that piecemeal rezoning should be subject to substantially stricter *procedural* requirements than comprehensive rezoning (507 P.2d at 30) because it is a reclassification of property and thus takes on the character of a quasi-judicial action (ibid., p. 26). Only true legislative actions, which are actions producing general policies or rules applicable to an open class of individuals, are granted a presumption of validity (ibid., p. 30). In jurisdictions that follow the *Fasano* rule, municipalities that enact piecemeal downzoning for solar access protection are subject to close judicial scrutiny for compliance with stringent procedural requirements.

out compensation. Downzoning that still permits development more intense than, or similar to, that permitted in nearby areas will generally be upheld.[21] If, however, an area is downzoned to an intensity level below that of the adjacent areas, some courts may strike down the rezoning as discriminatory or confiscatory.[22]

Political issues. Any sort of zoning change proposal will upset some property owners, who may feel they are being unfairly restricted for the benefit of others, and create a climate that might encourage lawsuits. Community opposition to downzoning in general, coupled with a lack of understanding about solar energy, may convince a municipality to limit (or ignore) the use of downzoning as a solar access protection tool. On the other hand, property owners who plan no further development of their own land, and who do not want neighbors to expand their structures either, would most likely welcome a downzoning to current heights or densities. Such situations often exist in suburbs. Similarly, urban redevelopment neighbors may seek downzoning.

Variances

The usefulness of zoning as a solar access protection tool may be weakened by the granting of variances—either from a solar access law or from a regular zoning law. Variances that allow higher structures, smaller setbacks, or different sizes or locations for accessory structures on particular lots could result in the shading of collectors. Even where solar access protection laws are adopted, the language of the law could fail to specify when variances will be denied or granted, creating uncertainty.

This variance problem can be avoided by enactment of an ordinance that prohibits the issuance of any type of development permit

21. *See*, e.g., Steel Hill Dev., Inc. v. Town of Sanbornton, 469 F.2d 956 (1st Cir. 1972) (downzoning conformed area to surrounding forest use); Coppolino v. County Bd. of Appeals, 23 Md. App. 358, 328 A.2d 55 (1974) (downzoning from high to low residential density conformed area to surrounding unimproved and single-family residential use); McGowan v. Cohalen, 41 N.Y.2d 434, 361 N.E.2d 1025 (1977) (downzoning from industrial to low-intensity residential conformed area to surrounding vicinity; valid notwithstanding 79 percent decline in market value of subject property).

22. *See*, e.g., Grimple Assocs. v. Cohalen, 41 N.Y.2d 431, 361 N.E.2d 1022 (1977); Aronovitz v. Metropolitan Dade County, 290 So. 2d 536 (Fla. Dist. Ct. App. 1974); Superior Uptown, Inc. v. City of Cleveland, 39 Ohio St. 2d 36, 313 N.E.2d 820 (1974); City of Cherry Hills v. Trans-Robles Corp., 181 Colo. 356, 509 P.2d 797 (1973); Obadash v. Mayor & Council of Dumont, 65 N.J. 115, 319 A.2d 712 (1974); Fred F. French Investing Co. v. City of New York, 39 N.Y.2d 587, 350 N.E.2d 381, *appeal dismissed*, 429 U.S. 990 (1976).

(including a variance) that would result in the shading of a protected collector. The only way a collector could then be shaded would be if the government compensated the collector owner for losses from the shading. Or the ordinance could prohibit the issuance of any development permit that would breach the regulations on heights and locations, with exemptions from these regulations available only in limited circumstances spelled out in the ordinance.

In the absence of legal provisions of this sort, variances will create public uncertainty and judicial headaches. Most state legislation delegating zoning power to local authorities includes the power to grant variances. The Standard Zoning Enabling Act also has such a clause. Because it is generally in the public interest for land to be productively used, ordinances typically say that a variance will be granted if an "unnecessary hardship" would result from strict enforcement of the zoning law. An unnecessary hardship occurs if, because of zoning, no feasible use can be made of the land.[23]

A variance will not be granted if the result would be detrimental to the public welfare. Courts will look at the impact on property values, the character of a neighborhood, traffic, and fire safety. This limitation is of interest to solar collector owners, even those in jurisdictions with no solar access laws, because it is possible that boards of appeals or adjustment (or the courts if a judicial review of administrative action is sought) could find it detrimental to public welfare to allow collectors to be shaded, and hence deny a variance.

If the statute or ordinance does not require it, adjacent landowners are not necessarily entitled to notice of a hearing on a request for a variance or special exception.[24] Not all zoning enabling acts

23. If a fair return on the land can be made without a variance, a variance cannot be expected even if a larger profit would result from its issuance. *See* Broadway, Laguna, Vallejo Ass'n v. Board of Permit Appeals, 66 Cal. 2d 767, 59 Cal. Rptr. 146, 427 P.2d 810 (1967). Economic hardships alone may not be sufficient grounds for a variance. ANDERSON, AMERICAN LAW OF ZONING, 3: § 18.22, p. 195 n.66. Court decisions also often explicitly require that the hardship be unique to that particular property, that is, that it not be a condition burdening other properties in the zone. Furthermore, the hardship must not be "self-imposed."

An example of how variances may work in a city with solar access protection follows: A city enacts an inadequate access law that does not make allowances for the slope of land. Lot A slopes steeply to the north, Lot B is on top of the hill to the south of Lot A. *Any* construction on Lot B would shade collectors on Lot A. The law has imposed a great hardship on the owner of Lot B, and a variance may well be granted to the owner of Lot B. In a city with a well written access law, variances would be granted only when their impact on access was negligible or clearly outweighed by a much greater public need.

24. ANDERSON, AMERICAN LAW OF ZONING, 3: § 20.19, p. 496.

contain public hearing and notice provisions for variances.[25] Because the great investment necessary to install a solar energy system could be rendered virtually worthless in the event of shading, states should at least require that notice and an opportunity for a hearing be given to the solar system owner.[26]

One approach to ensure solar system owners an opportunity to contest the granting of a variance or special exception would require each municipality to map the location of solar collectors for which building permits had been issued.[27] The issuing body would have to consult the map and determine whether it appears that a special exception or variance might result in impermissible shading.[28] If so, a hearing would be held to find out how much shading would actually occur. In the event of an impermissible amount, the request for a variance or special exception would be denied.[29] Such an approach has been incorporated in a comprehensive solar access bill introduced in Massachusetts. Both pieces of prototype legislation in this study contain similar provisions ensuring adequate notice and hearings. Giving notice and holding hearings will increase the workload of administrators. The expense and effort involved could be greatly reduced by clear standards as to when variances will be granted. Communities may shift some of this burden by requiring the variance seeker to give the notices.

Overlay Zoning

Through overlay zoning, property may be simultaneously located in two zones: a conventional zone and an overlay zone. The overlay zone imposes additional restrictions on property within its boundaries. The conditions of both zones must be satisfied before land may be altered, developed, or redeveloped.

Overlay zoning is primarily used when a special public interest is not fully compatible with the existing underlying zone. For example,

25. Ibid., § 20.17, p. 491.

26. A Virginia bill would prohibit the granting of a variance or special exception if the resulting construction would interfere with a solar collector's access to sunlight. Va. S.B. 854, H.B. 2014 (1977). If a variance or special exception were sought under this bill, it would be the responsibility of the collector owner to notify the locality of the probable conflict. However, a solar system owner would not even know that a variance or special exception had been sought.

27. Mass. S. 304, § 34 (1978).

28. Ibid., § 35(a).

29. Ibid., § 35(c).

overlay zones are used in San Francisco to preserve opportunities for views.[30] Either solar overlay zones or subcategories of existing zones should be considered as a way to avoid challenges under uniformity clauses in state constitutions.

Before designating solar access overlay zones or subzones, it will be necessary for a community to decide on the areas where it wishes to encourage the use of solar energy. The points raised earlier in discussing the role of comprehensive planning are relevant here. To be most effective as well as legally supportable, overlay zoning should be the result of detailed comprehensive planning. Based on extensive technical data and broadly based public policy, the plan should rationalize future decisions about the geographic location of proposed solar overlay districts. Since the overlay districts may not coincide with existing zoning districts, their geographic boundaries may be particularly difficult both to establish and to justify. While a comprehensive solar access protection plan cannot guarantee that an overlay system will be immune from legal attack, it should be the first step in implementing this kind of zoning.

Overlay zoning has been widely used at the local level.[31] It has been subjected to a few court challenges. In a Wisconsin contract zoning case, the court said *in dictum*:

> Where the imposition of conditions on land development is desirable, it might better be done by uniform ordinances providing for special uses, special exceptions and overlaid districts.[32]

In a Maryland case, the court said of overlay zoning: "The concept has the desirable characteristics of being a 'comprehensive rezoning' as it relates to one particular use necessary but special in character."[33]

To implement solar access overlay zones, a local government must insert new regulations in the zoning text. The new regulations should delineate restrictions such as those on the height and placement of buildings and vegetation that could unreasonably inhibit

30. San Francisco, Cal., Municipal Code pt. II, ord. 234–72, art. 2.5 (1972) (height and bulk districts).

31. *See* Musego, Wis., Zoning Ordinance § 9 (1963); Boulder, Colo., Zoning Ordinance art. II, § 37-201(a) (Ord. 3750, 1971); Baltimore County, Md., Zoning Ordinance § 405 *et seq.*, as amended by bill no. 40 (1967); Portland, Ore., Zoning Ordinance ch. 33.62 *et seq.* (1972); San Francisco, Cal., Municipal Code pt. II, ord. 234–72, art. 2.5 (1972).

32. State ex rel. Schimenz, 174 N.W.2d 533, 539 (1970).

33. Swarthmore Co. v. Kaestner, 266 A.2d 341, 344 (Md. Ct. App. 1970).

solar access. After the local government has defined and mapped the affected zones, any person wishing to develop, redevelop, or modify an existing property in that area would be required to satisfy the new solar restrictions as well as those of the underlying zone (which would not be linked directly to solar access).

A municipality could require a site plan review to ensure that conditions of the solar overlay would be achieved. Considerations in a site plan review process often include:

- □ Setbacks of buildings from surrounding property
- □ Existing and planned vegetation and landscaping
- □ Slopes and drainage patterns
- □ Joint use of parking facilities
- □ Storm water retention and erosion provisions

Solar access protection could easily be added to this list of concerns. By incorporating site plan review procedures into overlay zoning, a local community would gain a means of dealing with vegetation. As there are major redevelopment and rehabilitation efforts underway in many American cities, opportunities exist to apply site plan review techniques for solar access purposes.

Overlay zoning and the techniques that may be linked to it (such as site plan review) offer a promising method of protecting access to sunlight. The specific restrictions imposed through overlay zoning may be nontraditional. For example, solar envelopes (Chapter 5) or collector recordations (Chapter 7) might both employ overlay zones.

5
Solar Envelopes

Solar envelopes are basically just zoning height limitations that are established with the changing daily and seasonal positions of the sun in mind, and that may vary over different parts of a lot.[1] They apply to both structures and vegetation—to the "growable" as well as the "buildable" areas of lots. In developed areas, solar envelopes would most likely be applied through the use of overlay zones.

Traditional zoning regulations define an "envelope" inside which development is permitted. Typically, however, the envelope is nothing more than a simple rectangular box fitting just over the buildable area of the lot (see Figure 5-1). To further meet some of the purposes of height controls—particularly the provision of adequate light and air and the preservation of views—several communities have chosen a different shape for their height envelopes.[2] Rather

1. Cities with no zoning, such as Houston, may be able to achieve almost the same result as solar envelopes through modification of height regulations in building codes.

2. NORMAN WILLIAMS, JR., AMERICAN LAND PLANNING LAW, 5 vols. (Chicago: Callaghan & Co., 1975), 1: §§ 35.07–35.11, pp. 668–72, lists among others, New York City; Philadelphia, Pennsylvania; Denver, Colorado; and Seattle, Washington. Others include Fairfax County, Virginia; Albuquerque, New Mexico; and Tisbury Island Road District, Massachusetts. *See* AMERICAN PLANNING ASSOCIATION, PROTECTING SOLAR ACCESS FOR RESIDENTIAL DEVELOPMENTS: A GUIDEBOOK FOR PLANNING OFFICIALS (Washington, D.C.: U.S. Government Printing Office, 1979), for a discussion and illustrations. A very similar result has been achieved in Japan under the "sunshine" amendments added to the Japanese national Building Standard Law (BSL) in the 1970s (KENCHIKU KIJUN HO, Law No. 201 of 1950, as amended). Although these amendments are written in terms of shadow controls,

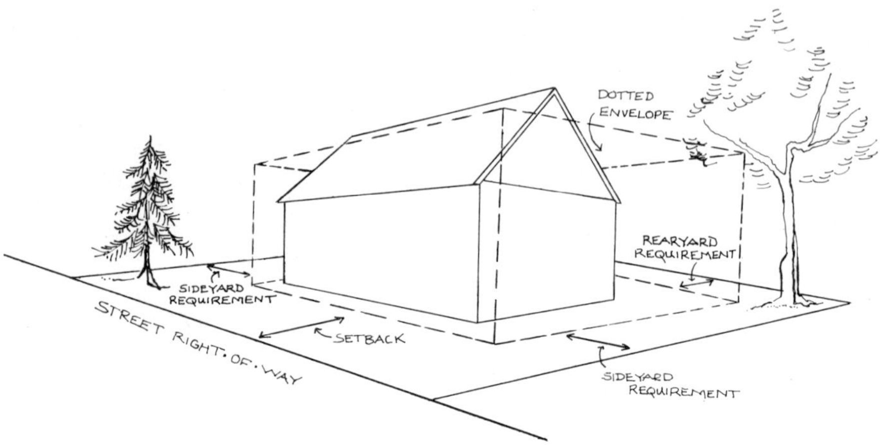

FIGURE 5-1
THE ENVELOPE FOR CONVENTIONAL ZONING

Height is usually measured half-way up a sloped roof, so half of roof may penetrate the top of the envelope. Vegetation may be outside the envelope.

than a flat plane parallel to the ground, the top of the "box" or envelope consists of one or more planes sloping at various angles (see Figure 5-2).

In fact, the original comprehensive zoning ordinance in this country, the 1916 New York City ordinance, used this kind of height control.[3] These "bulk plane ordinances" or "light angle controls" are intended to provide light and air for more pleasant and healthful living conditions. Although not used at present for the express pur-

rather than solar envelopes, they mean that development must fit within sloping planes. The result of this approach is evident (as in New York City) in the tiered, angled Tokyo skyline. Further discussion of Japanese legal approaches to allocating sunlight can be found in Gail Feingold Takagi, *Designs on Sunshine: Solar Access in the United States and Japan,* CONNECTICUT LAW REVIEW 10 (1978): 123; and *Tokushu Nisshoken* [Special Issue on Sunshine Rights], JURISATO [JURIST], a Japanese legal periodical published in Tokyo by Yuhikaku Ltd., January 1974 (not available in English). Appendix A outlines a method for developing the heights and slopes of the planes that make up a solar envelope. Another method is outlined in APA's GUIDEBOOK FOR PLANNING OFFICIALS.

3. WILLIAMS, AMERICAN LAND PLANNING LAW, 1: § 35.09, pp. 668–70. The result of those controls was the familiar wedding cake shape of many of the city's early skyscrapers. That result can be avoided by having planes nearer to the horizontal that begin at some height above the ground, instead of at nearly vertical angles and beginning at the ground as in the New York City ordinance.

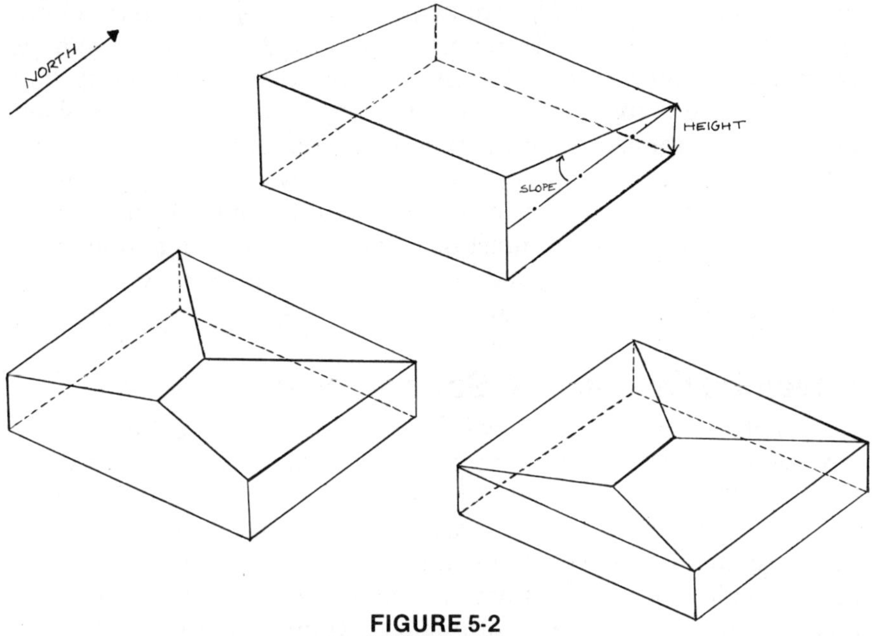

FIGURE 5-2
ENVELOPE SHAPES

pose of assuring access to sunlight for solar collectors,[4] they could easily be adapted to do so. The prototype ordinance at the end of this chapter is based on this approach, and suggests some types of envelopes that could be used.

The size of solar envelopes can vary with the size, shape, slope, and orientation of lots. Rules governing the envelope applicable to a particular lot should apply uniformly to identical lots within that overlay zone, but envelopes can vary from zone to zone depending on the existing or planned development in the zone and the extent of solar access protection to be given. A solar envelope over lot A will protect solar access on nearby lots. Envelopes over nearby lots will, in turn, protect A's access.

Solar architect Ralph Knowles, of the University of Southern California, suggests an envelope definition that results in comparatively small envelopes. Knowles' work has focused more on new developments than on additions to existing structures and vegetation. His envelopes are very low tent shapes that optimally keep all shad-

4. Except in Albuquerque, New Mexico. But the light planes there are 45 degrees and 60 degrees, which provide only incidental solar access protection during the heating season.

ing within the bounds of the lot in question during the relevant hours (for example, in winter from about 9 A.M. to 3 P.M. and in summer from about 6:30 A.M. to 5:30 P.M. for Los Angeles.[5] While such access may be possible or desirable in a new development, it is wholly unrealistic if applied to an existing developed area. Even in new developments, such solar access would generally be unnecessary unless ground-level collectors are located right on lot lines. In most developed areas, the envelopes must be much larger than in new areas because more shading beyond lot boundaries already occurs than under ideal solar access conditions.

Envelope Options for Solar Access

There are at least three ways that municipalities could use solar envelopes to protect solar access. The first option is to redefine height so that the legal height of a building corresponds to its highest part, to limit exceptions from height regulations, and to extend the height regulations to vegetation anywhere on the property.[6] Overlay zones would not be necessary; the amendment could be made directly to the regular zoning ordinance. Allowed densities would not change, only the roof shapes would be different. This would be equivalent to adopting a simple flat-topped rectangular envelope over the whole property (see Figure 5-3). (Alternatively, only vegetation on the northern part of lots could be regulated.)

This option might be politically feasible in some areas where more complex solutions, such as the prototype legislation in this chapter, would not be. Although a simple-to-draft change, this option would offer solar access protection superior to that provided by existing land use controls. Structures (and perhaps trees) that already exceed the redefined height limits would presumably be exempted from the law rather than made nonconforming uses. Lacking refined enforcement and remedy provisions, this approach would be less effective than some legislation, but it would fare rather well when measured against such criteria as compatibility with existing laws (even granting the slight risk of invalidation if there is not state-enabling legislation for vegetation controls) and with other

5. Figures from Ralph Knowles, *Solar Ethics—Urban Form,* in PASSIVE SOLAR: STATE OF THE ART, ed. Don Prowler, 3 vols. (Newark, Del.: American Section of the International Solar Energy Society, 1978), 3: 843, 846. *See also* RALPH KNOWLES, ENERGY AND FORM (Cambridge, Mass: MIT Press, 1974).

6. *See* Chapter 4.

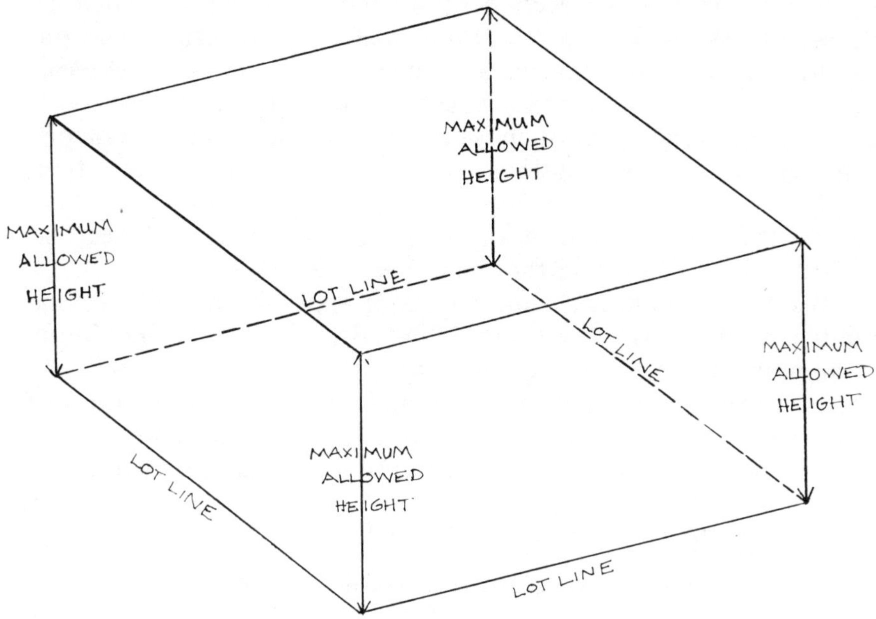

FIGURE 5-3
SIMPLE ENVELOPE
No land slope is assumed.

public policies. There would be little cost to communities either in terms of implementation or development restraint. Burdens would be clearly and fairly allocated. For these reasons, this option is more effective than some of the other "simple" proposals such as those bills based on water law analogy and nuisance.

An obvious limitation to this option is that generally only rooftop access would be assured. It is also inappropriate for areas with many trees taller than existing structure height limits. It would be politically difficult to require such trees to be cut, even where it would not result in a net energy loss because of the cooling and windbreak functions of trees. A variation of this option would be to allow tall vegetation to remain until a collector was installed and was actually shaded.

The second option is like the first but uses rezoning to lower heights or densities uniformly. The planes formed by height limits would still be parallel to the ground—i.e., the envelopes would be box-shaped. This approach gains effectiveness at the expense of political acceptability; implementation would also be more difficult.

Either the first or second option could be strengthened by adopting special ordinances to deal in greater detail with vegetation, grandfathering, maintenance, and enforcement provisions. However, a comprehensive approach such as the prototype legislation is probably preferable to such complex tinkering with controls. Although the first two options would guarantee more access than present laws do, their flat-topped box-like shapes do not account for the movements of the sun. In some instances, they would protect too much access; in other cases they would provide too little.

The third option is a comprehensive approach where envelope zoning is applied to both vegetation and structures, and height limits are sloping planes. The size and shape of the envelopes could be tailored to the actual environmental and developmental conditions in various areas of a municipality. For localities that need comprehensive solar access protection, this option is recommended as the best overall way of protecting solar access.

Components of this third option could include:

- A comprehensive plan element for solar access
- Careful mapping specifically for solar access (as opposed to the usual mapping which largely ignores such factors as existing vegetation and topography)
- Procedural safeguards for all parties
- Provision for exceptions to prevent undue hardship

In spite of the considerable data gathering and other efforts that must go into their design, solar envelopes appear to be the best presently available legal device for protecting solar access for entire cities or neighborhoods. They can prevent existing solar access from being obliterated by reckless development and are flexible yet comprehensive. Careful advance planning discourages later litigation and lessens the probability of conflicts between individuals. Furthermore, envelopes require only a small modification of familiar laws and can place the fewest possible restrictions on development consistent with the preservation of future access. The inherent flexibility of solar envelopes is one of their major attractions. It is also a drawback, because of the planning required. A recordation approach (Chapter 7) may allow greater flexibility on a case-by-case basis, but there is less promise of clear compliance with an overall plan and no provision for future collector installations. The prototype here provides for revisions to the solar zone map and solar envelopes as needed, allowing flexibility through time while keeping the same basic administrative and enforcement structures.

Readers are urged to study the comment sections of the following prototype legislation. Some of the points and analyses in the comments do not appear elsewhere in this report.

An Ordinance Concerning Solar Energy, Providing for Solar Access Protection Through Solar Envelopes

Introductory comment. The following prototype ordinance is one of many possible implementations of solar access protection by zoning. For the reasons discussed in the sections in Chapter 4 dealing with the use of overlay zoning and rezoning, this ordinance imposes overlay zones with solar envelopes for each lot. These envelopes consist of planes of various heights and slopes, depending on the particular overlay zone where the property lies. (Heights and slopes can be chosen with the aid of Chapter 2 and Appendix A.) Structures and vegetation are prohibited from "breaching" (sticking through) the envelopes unless a general exemption applies or an exemption permit has been granted under a special procedure (see Figure 5-4).

The key provisions of this ordinance are the two sections that provide for the effect and enforcement of the envelopes (Section 6) and the general exemptions and procedures for application for special exemption permits (Section 7). They include most of the substantive matter in the ordinance.

The effect and enforcement provisions, as in the prototype collector recordation ordinance in Chapter 7, provide separately for structures and vegetation. City officials would be prohibited from issuing building permits, zoning permits, variances, special permits, and other "development permits" for structures if the structures would breach the solar envelope planes over the lot in question, unless an exemption applied (Section 6(a)). A "solar access beneficiary" (a person whose property would be—or is—shaded by the breaching part of the structure) may appeal the issuance of such a

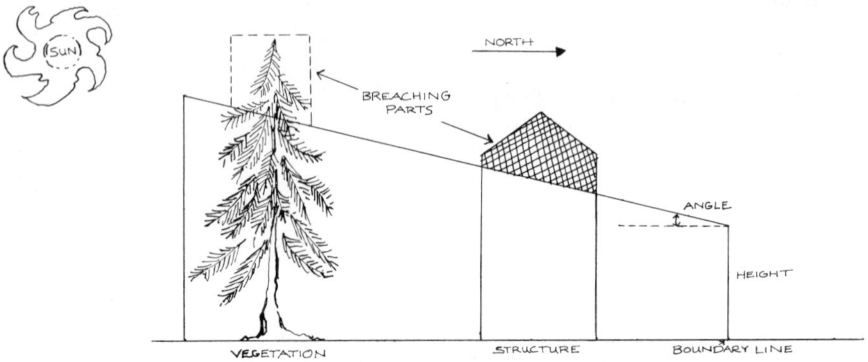

FIGURE 5-4
BREACH OF SOLAR ENVELOPE

permit. He could receive an injunction against the construction, or an order for removal of the breaching part (if feasible), or damages (where removal is infeasible). He is also entitled to other appropriate damages, court costs, and attorney's fees (Sections 6(a) and (b)).

New vegetation and vegetation growth that breach an envelope are prohibited, and an obligation to trim vegetation at least once a year is imposed (Section 6(c)). Alternative provision is made for either exempting all growth of existing vegetation or exempting only the size of existing vegetation at the time the overlay zone is adopted. Elaborate provisions are made for giving notice of vegetation breaches, court orders for removal, and, if necessary, court orders allowing the city to enter to remove breaching vegetation, with costs paid by the owner of the breaching vegetation (Section 6(d)–(f)). Solar access beneficiaries can agree in writing to permit breaching vegetation to cast shadows on their land (Section 6(h)). Any party required to trim is entitled to reimbursement from the beneficiary where the vegetation can be shown to have been in place before the overlay zone was imposed (Section 6(i)).

There are general exemptions for breaches by utility poles, wires, flagpoles, antennas, and such other insubstantial breaches as are provided for in regulations (Section 7(a)). Exemption permits are available for either structures or vegetation where the applicant can show either that he cannot achieve the underlying density provided in the current zoning without breaching the envelope or that the proposed structure or vegetation would not prejudice the interests of any solar beneficiary (Section 7(b)–(d)).

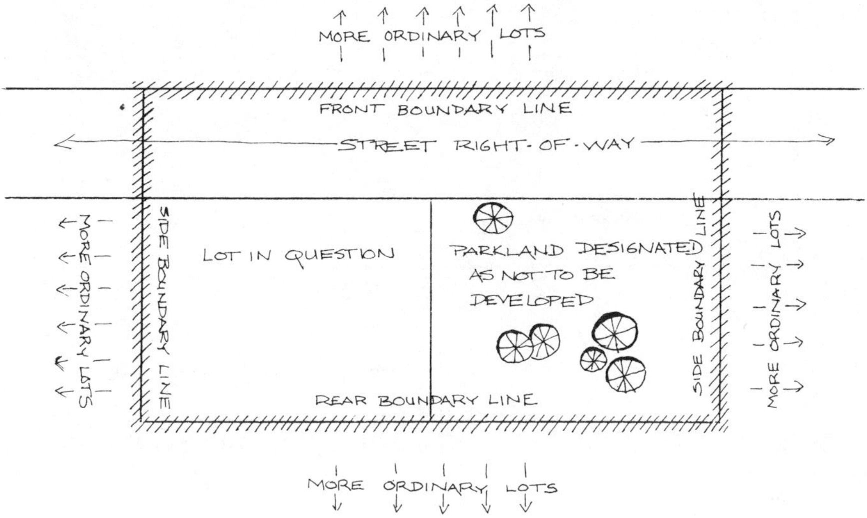

FIGURE 5-5
BOUNDARY LINES

The definitions (Section 2) and envelope construction (Section 5) may initially appear strange and possibly confusing compared to usual land use terms and rules, but their logic should become apparent.

"Boundary lines" (see Figure 5-5) are not the usual lot lines but enclose a larger area, such as adjacent street right-of-ways and undeveloped parklands. Such definitions are more liberal toward development than the usual lot lines of the burdened property owner because some lands are not likely to require protection against shading.

"Boundary base lines" and "base plane" (see Figure 5-6), in the simple case of a lot that is a smooth plane, correspond to the boundary lines and the plane of the lot. Where the terrain is uneven, the averages along the lot lines, rather than the average grade over the whole lot, are used for constructing the envelopes. There is thus a penalty for lots that rise toward the center and a bonus on lots that dip toward the center. The result is that the solar access for nearby lots will be roughly constant, despite the nature of the terrain on the burdened lot.

"Height" of envelope planes is measured vertically in all cases (see Figure 5-7). "Bearing of a boundary line," and the "northern," "southern," "eastern," and "western boundary lines" (see Figures

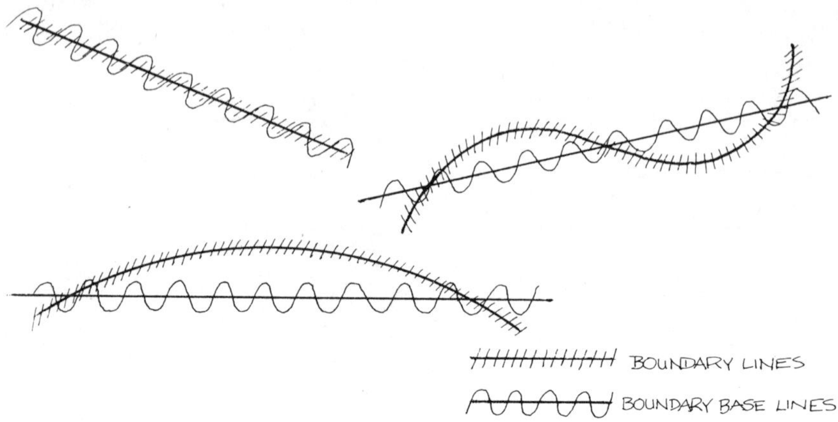

FIGURE 5-6
SIDE VIEWS OF BOUNDARY LINES AND BOUNDARY BASE LINES

Boundary base lines are straight-line averages of the boundary lines, which go up and down with the terrain.

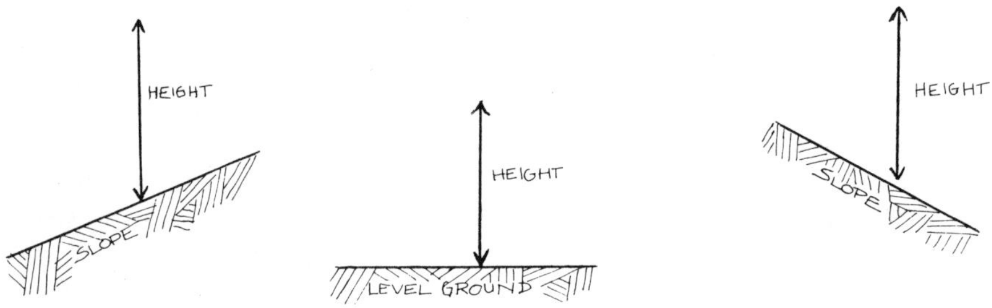

FIGURE 5-7
HEIGHT

Height is always measured vertically.

5-8 through 5-12) place every boundary line in one of these categories so that a unique height and envelope plane slope for the envelope plane applies. Under this provision it is possible for oddly shaped lots to have more than one (or no) boundary line for any particular direction (see Figure 5-13).

Some solar access legislation seems to assume that all properties have a northern or southern property line that runs east and west. In such legislation it is not clear what should be done with a property line that runs, for example, from northeast to southwest. Categorizing boundary lines in the way the prototype legislation does allows more severe restraints on development toward the northern end of

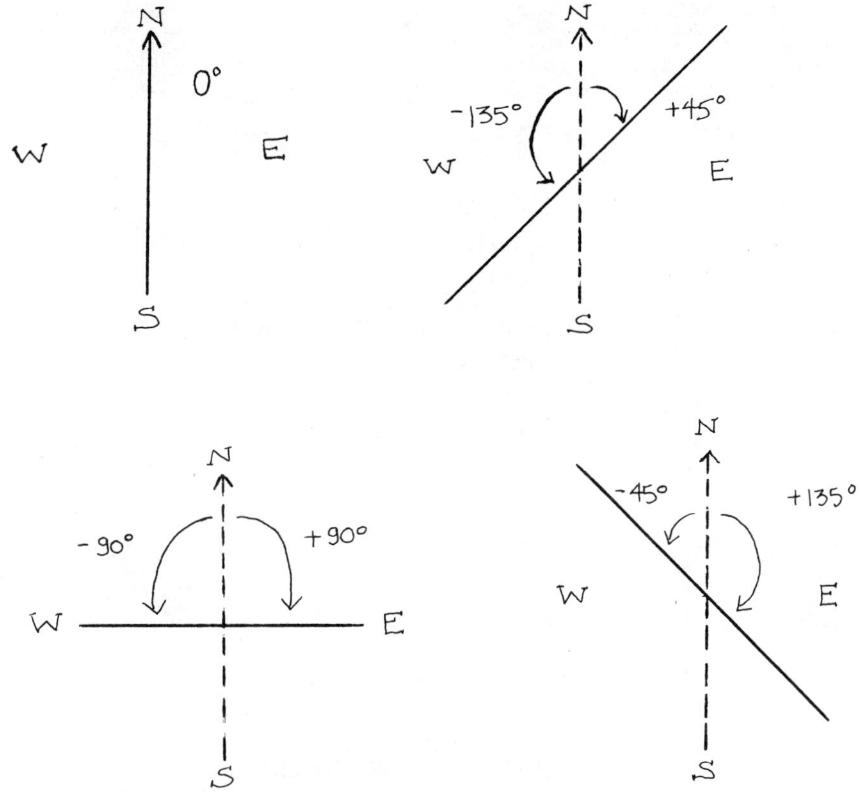

FIGURE 5-8
BEARING OF A BOUNDARY LINE

View is looking down from above.

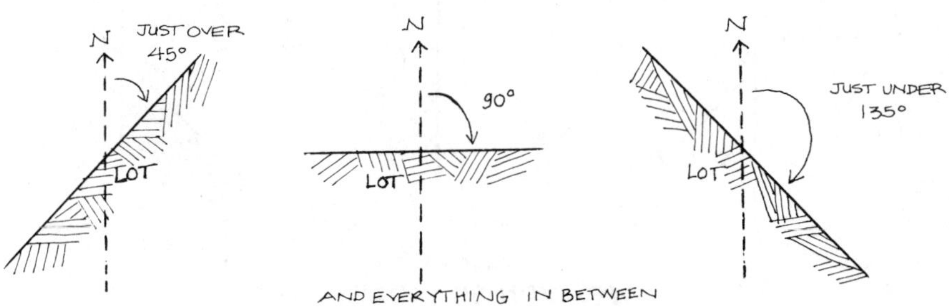

FIGURE 5-9
NORTHERN BOUNDARY LINES

View is looking down from above.

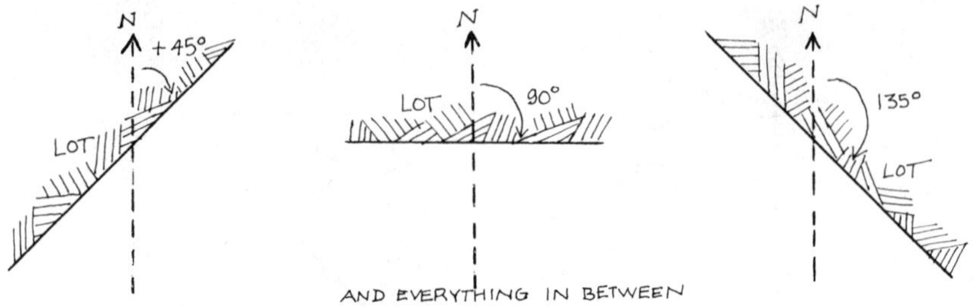

FIGURE 5-10
SOUTHERN BOUNDARY LINES
View is looking down from above.

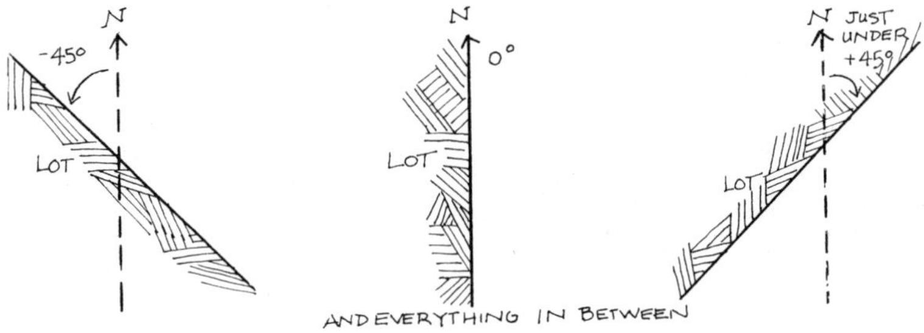

FIGURE 5-11
EASTERN BOUNDARY LINES
View is looking down from above.

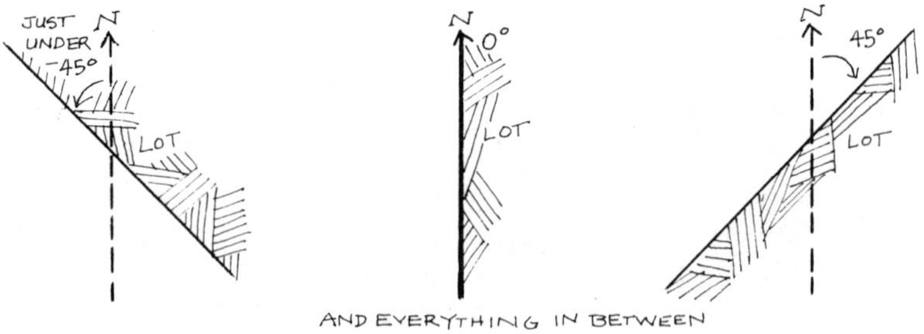

FIGURE 5-12
WESTERN BOUNDARY LINES
View is looking down from above.

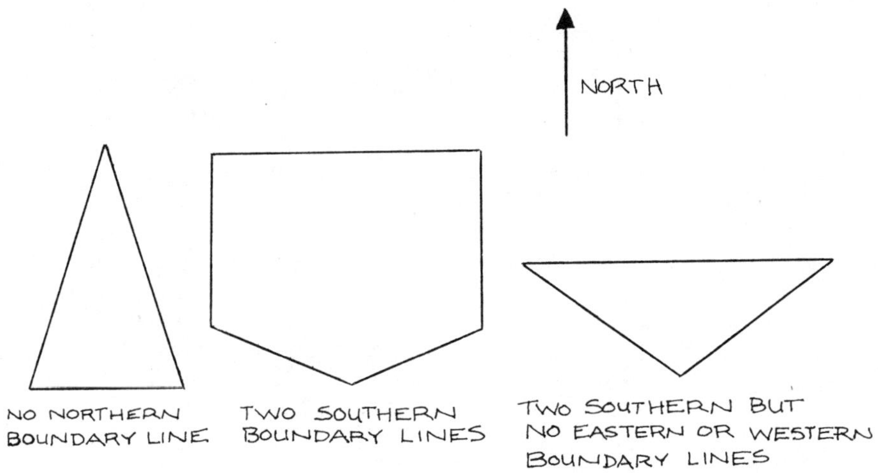

FIGURE 5-13
ODD BOUNDARY CASES

lots and less severe restraints toward the east and west sides to protect early morning and late afternoon sun for winter solar heating and year-round hot water. Restraints toward east and west sides and even toward the southern ends of lots will be necessary for efficient use of summertime active solar air conditioning and photovoltaics, because the sun rises north of east and sets north of west during the summer.

Four types of overlay zones are described in the ordinance (not all may be needed in any given city). Each type of zone calls for a particular shape for solar envelopes. The size of envelopes, however, can vary within each zone depending on the size of the lot.

Envelope shapes can be constructed in one of three ways: the simplest method is to draw a single plane parallel to the base plane at the height specified for the particular overlay zone (see Figure 5-14). This is roughly comparable to the usual height limitation in zoning, but this control applies to vegetation as well. A second, somewhat more complex approach, is to draw a plane beginning at the particular edge of the lot at a height specified for the zone and sloping up over the lot at a slope measured in relation to the horizontal, specified for the particular overlay zone (see Figure 5-15).

The third possible method is the same as the second except that the angle is measured relative to the horizontal if the lot is level or if it slopes down to the south. However, if the lot slopes down to the north, the envelope plane is measured relative to the base plane (see

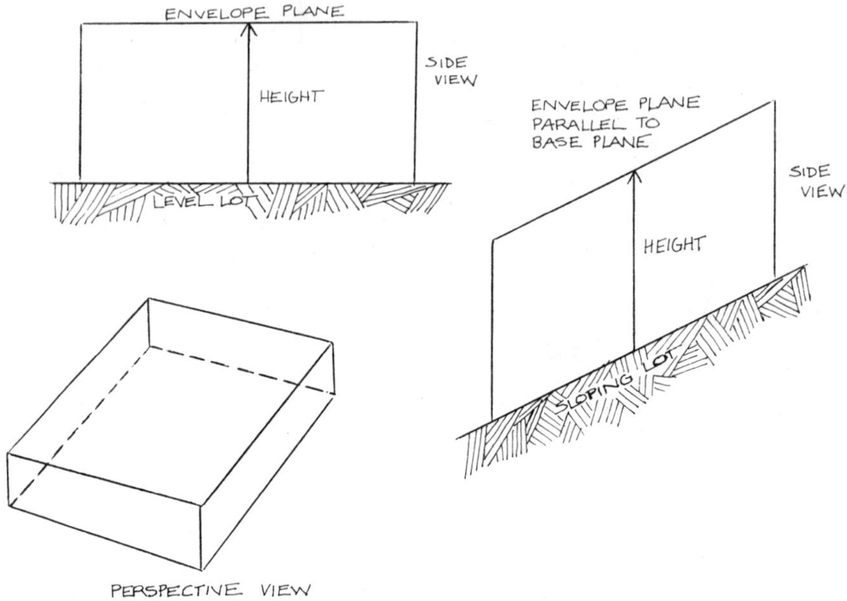

FIGURE 5-14
ENVELOPE IN TYPE 1 OVERLAY ZONE WITH PLANE CONSTRUCTED BY THE FIRST METHOD

Figures 5-16 and 5-17). This allows for fewer restrictions on the development of slopes, but permits the loss of significant solar access for the adjacent property. This is a reasonable tradeoff where development is to be fairly dense on all slopes. (For the best solar access protection, envelopes should be drawn using the second method, but the ordinance as written uses only the first and third methods because these would be more widely applicable.)

In this prototype legislation, the Type 1 overlay zone has envelopes with planes drawn by the first method. The Type 2 overlay zone has envelopes with planes drawn by the third method on northern boundary lines. The Type 3 overlay zone adds more planes drawn by the third method on the east and west boundary lines, and the Type 4 zone adds even more planes drawn by the third method on southern boundary lines.

Even the simple envelopes in Type 1 overlay zones would provide solar access protection far greater than exists under the usual land use controls. Type 2 controls, with sloping planes on northern boundary lines, could provide better solar access with less overall restraint on development (restraints would be greater on northern

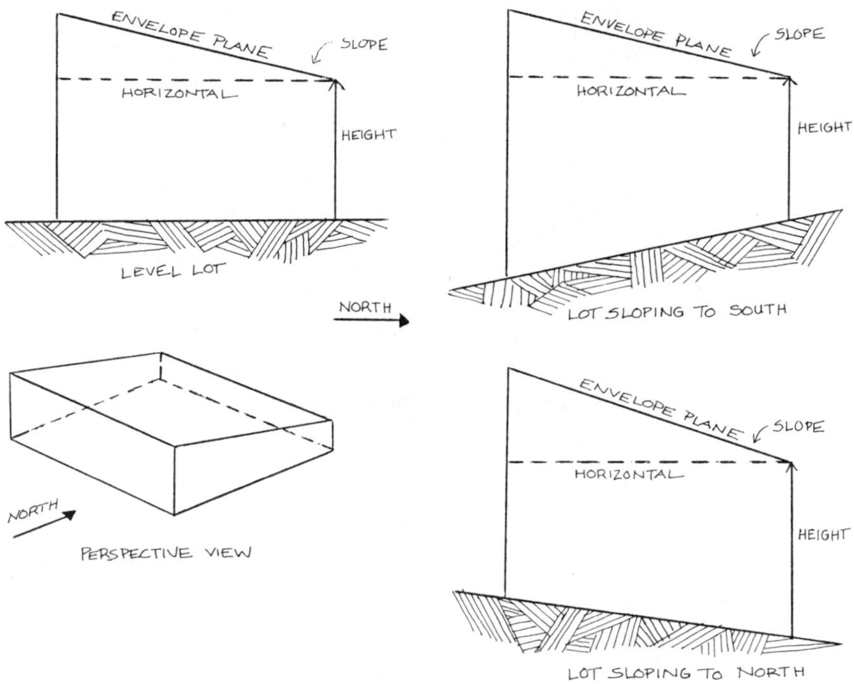

FIGURE 5-15
ENVELOPE WITH PLANES CONSTRUCTED BY SECOND METHOD

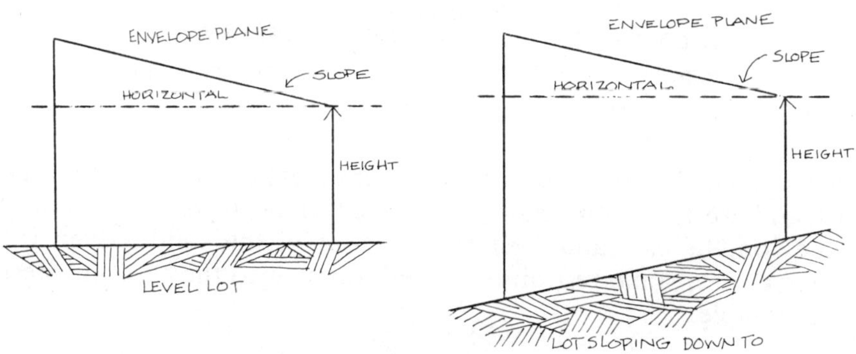

FIGURE 5-16
ENVELOPE IN TYPE 2 OVERLAY ZONE WITH PLANES CONSTRUCTED BY THIRD METHOD

Note that plane slopes for Types 3 (with added planes on east and west) and 4 (with added planes on east, west, and south) also use this method of measurement.

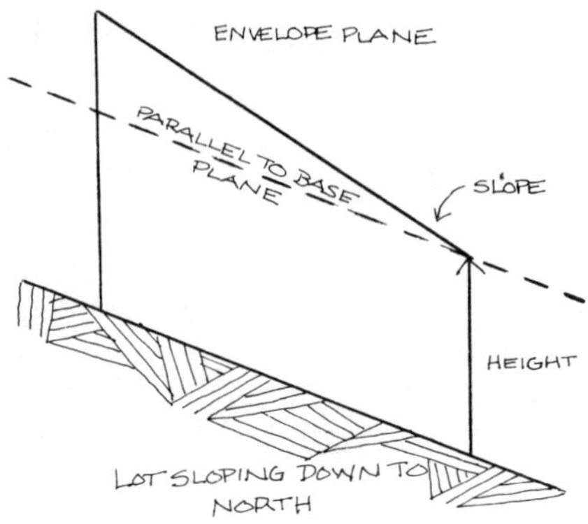

FIGURE 5-16 (continued)

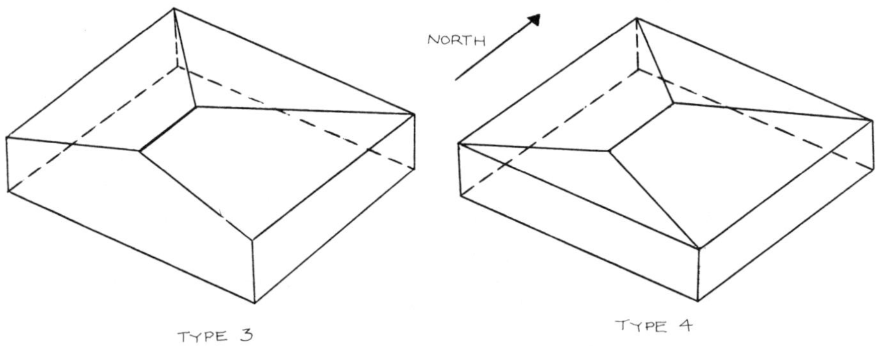

FIGURE 5-17
ENVELOPES IN TYPE 3 AND TYPE 4 OVERLAY ZONES

These are constructed as in Type 2, but with added planes on east and west (Type 3) or east, west, and south (Type 4).

ends of lots). For protection of summer access to sunlight from the east and west, it is necessary to use envelope types with sloping planes on the east and west and, desirably, the south. Even for winter access, some restraints on development to the east and west are advisable.

Note: This prototype ordinance is not intended to be adopted in its present form. It is a longer law than will be required in many localities, and is deliberately comprehensive to serve as a reminder of the many possible issues that may be encountered. Many of the goals expressed in this ordinance could, instead, be included in administrative regulations. Much of the language could also be used in a state-level law or guidelines accompanying such a law. The ordinance format was selected because of its general appropriateness for solar access legislation. When brackets appear in the legislation they indicate either (1) alternative approaches, or (2) places where local expertise must supply the appropriate name or number. Numbers inside brackets are simply suggestions of what may be reasonable.

Enactment Clause with Purpose and Findings (Alternative One)

The City Council of [city name] finds that:

(a) The energy supplies of the City are dependent on suppliers located beyond its legal jurisdiction, thus beyond the control of its citizens.

(b) The public health, safety, and welfare of the citizens of this City are dependent upon reliable and affordable energy sources for such applications as heating and cooling of homes, offices, industrial plants, and commercial facilities; for hot water for these facilities; and for electric power. Any interruption of supplies affecting the heating and cooling of buildings, the supply of hot water, or the supply of electricity could cause irreparable harm to individuals and to the public welfare.

(c) The nation's energy supplies are diminishing and the long-term trend is toward depletion of the world's fossil fuel resources, with the possibility of shortfalls in energy needed to protect and promote the public welfare of the citizens of this City.

(d) The price paid by citizens for all forms of energy is rapidly increasing; price increases will continue for reasons beyond the ability of the City or of individual citizens to control adequately.

(e) Solar heating and cooling of buildings, solar-heated hot water, and solar-generated electricity can provide a significant contribution to the energy demands of this City. Additional energy can be saved by using solar energy for industrial and commercial processes.

(f) The use of solar energy systems reduces air and water pollution resulting from the use of conventional energy sources and reduces requirements for capital, land, water, and other resources needed for central-station generation of electricity.

(g) All citizens of this City will thus benefit from the increased use of solar energy.

(h) A significant barrier to the rapid conversion of this City's heating, cooling, hot water, and electrical energy loads from vulnerable, expensive energy sources to solar energy is the concern among citizens that the solar radiation necessary for solar collectors to function will be blocked by structures or vegetation located on property under the control of others, including the City. Existing laws and regulations are inadequate to provide the degree of protection that is needed.

(i) The provisions of [state enabling legislation that grants zoning and/or home rule powers to the city] grant to this City the power to regulate the development of property for the purposes of protecting the health, welfare, and safety of its citizens. Regulation of structures and vegetation on property, including City-owned or controlled property, is necessary to ensure access to energy from the sun, and therefore such regulation serves to protect the health, welfare, and safety of the citizens.

(j) [Recite other state actions (such as state tax incentive legislation) showing that it is the policy of the state to expedite solar energy use and to broaden zoning powers for environmental protection purposes and to meet new needs in housing and other areas as they arise.]

(k) Pursuant to the power granted to this City to regulate land development, and for the purpose of promoting speedy conversion to nonvulnerable energy resources using solar energy, the City Council of [city name] does hereby enact the following Ordinance:

Comment on Enactment Clause, Alternative One. This kind of extensive enactment clause is recommended where the state legislature has granted a city only the ordinary powers to zone (as under, for example, the Standard State Zoning Enabling Act [SZEA]). The power to zone in order to regulate height and placement of structures is sufficiently well established in American law so that the extensive legislative finding contained in this Enactment Clause may seem unnecessary. Existing "light and air" laws are based mainly on health protection purposes, however, and not on the power to regulate for the general welfare. Because solar access protection laws will be justified mainly on the latter ground, it is wise to include detailed findings.

Furthermore, this ordinance imposes height limitations on vegetation, a power that is not expressly granted by the SZEA. Vegetation control for the purpose of protecting access to sunlight may be within the police power of the city under zoning or home rule provisions, but the legal framework is, as yet, undeveloped and uncertain. The extensive recitation of the factual background that underlies the ordinance is mainly for the purpose of providing a court, in the event of a challenge, with sufficient information to judge the public purpose—and hence the validity under the police power—of the imposition of vegetation controls.

Enactment Clause (Alternative Two)

The City Council of [city name], pursuant to the authority granted to it by [title and section number of state statute that specifically authorizes this ordinance], hereby enacts the following Ordinance:

Comment on Enactment Clause, Alternative Two. This alternative is for use in states that have already enacted legislation enabling, encouraging, or requiring localities to protect solar access. When such legislation exists, there may be no need for the extensive findings in Alternative One. More elaborate findings, such as those in the first alternative, may nonetheless strengthen the ordinance against attack.

SECTION 1 Short Title

This Ordinance may be cited as the "[city name] Solar Access Ordinance."

SECTION 2 Definitions

As used in this Ordinance, the following terms shall have the meanings shown:

(a) "Boundary line." A boundary line, running along any street right-of-way or street easement, is ordinarily along the property line of the lot on the far side of the street right-of-way or easement. Any other boundary line is ordinarily the property line between the lot concerned and the adjoining property. Where an adjoining property along what would ordinarily be the boundary line belongs to the City, is not developed, and has been designated as land that will not be developed on the map prepared and adopted under Section 4 of this Ordinance, the boundary line is treated as running along the property line on the far side of that adjoining lot. [See Figure 5-5.]

(b) "Boundary base lines." If a particular boundary line is on level ground or on uniformly sloping ground, the boundary line is considered the base line. If the ground surface along the boundary line is irregular, the boundary base line is considered established at the average slope and average height of the ground along the boundary line. [See Figure 5-6.]

(c) "Base plane." If the ground surface of a lot as defined by its boundary lines is flat, that plane is considered the base plane. If the ground surface is irregular, the base plane is constructed as the plane that is the closest fit to the boundary base lines.

(d) "Height." In relation to a solar access envelope plane, height of a point is the vertical distance from the boundary base line or base plane to that point. [See Figure 5-7.]

(e) "Bearing of a boundary line." A boundary line that runs due north or south has a bearing of zero degrees. Bearing of any other boundary line is its angle as measured from due north positively clockwise (turning from north toward the east) or negatively counter clockwise (turning from the north toward the west). [See Figure 5-8.]

(f) "Northern boundary line." A boundary line where the bearing of the boundary line is greater than 45 degrees and less than 135 degrees and the property lies to the south of the boundary line. [See Figure 5-9.]

(g) "Southern boundary line." A boundary line where the bearing of the boundary line is greater than or equal to 45 degrees and less than or equal to 135 degrees and the property lies north of the boundary line. [See Figure 5-10.]

(h) "Eastern boundary line." A boundary line where the bearing of the boundary line is greater than or equal to minus 45 degrees and less than plus 45 degrees and the property lies to the west of the boundary line. [See Figure 5-11.]

(i) "Western boundary line." A boundary line where the bearing of the boundary line is greater than minus 45 degrees and less than or equal to plus 45 degrees and the property lies to the east of the boundary line. [See Figure 5-12.]

(j) "Breaching the solar access envelope." A structure or vegetation breaches a solar access envelope if any part of it lies above the level of solar access envelope plane that applies at that point over the property in question. [See Figure 5-4.] Such a structure or vegetation is called a breaching structure or breaching vegetation. The part that lies above the envelope plane is called the breaching part. The party whose structure or vegetation breaches the envelope is called the breaching party.

(k) "Solar access beneficiary." Any person or persons on whose property shadows are cast or will be cast by the breaching part of a particular structure, proposed structure, vegetation, or proposed vegetation.

(*l*) "Development permit." Any permit or authorization issued by the City as a prerequisite for undertaking any development. Permits and authorizations customarily known as building permits, zoning or rezoning permits, variances, special permits, plat approvals, or subdivision or large-scale development permits are included.

(m) "Direct solar energy." Direct radiant energy received from the sun.

(n) "Solar collector." A device, or combination of devices, structures, or parts of a device or structure that transform direct solar energy into thermal, mechanical, chemical, or electrical energy and that contribute significantly to a structure's energy supply.

Comment on definitions. The particular forms of envelopes generated by these definitions (and under Section 5) represent just one possibility among many. This system allows for a simple set of envelopes at the outset which could be developed into more complex envelopes later, if the necessary technical and funding assistance for surveys became available from the states or federal government (see Chapter 12).

The definitions relating to envelopes have already been discussed in the introductory comment. The definitions (j), (k), and (l) for breaching, solar access beneficiary, and development permit have been introduced here to avoid wordiness in the substantive provisions.

The solar energy and collector definitions, (m) and (n), are necessary only for the part of the exemption procedure in which the applicant may show that the use of solar collectors is impractical in any location that would be shaded by the proposed structure or vegetation (Section 7(d)(1)(C)). For all other purposes, collectors can be ignored. Solar collectors for swimming pool heating are included only where pools are treated as structures under local building and zoning codes. If they are not covered and such applications of solar energy are likely, they can be added to the collector definition.

SECTION 3 Applicability of Ordinance

(a) Governmental organizations not under the jurisdiction of the City (such as the State and Federal governments) may elect to enjoy the benefits of solar access under this Ordinance if they also consent to be bound by its restrictions. An organization not having so consented may neither protest breaches of a solar access overlay under Section 6 nor protest an exemption application under Section 7.

(b) City property shall be subject to this Ordinance. The City may protest the breach of a solar envelope under Section 6 or an exemption application under Section 7. In such actions the City shall be represented by [appropriate agency].

(c) When the provisions of this Ordinance conflict with any other laws or regulations relating to land use, the requirements that most restrict development shall apply.

Comment on applicability. States and the federal government should be encouraged to consent to be bound by solar access legislation. Section 3 provides an incentive for states and the federal government to follow local land use rules protecting solar access. The ordinance makes it clear that its benefits go only to those who will be bound by it.

Subsection (b) allows the city on its own motion to act to protest the breach of a solar envelope or any exemption application. It may wish to do so to protect neighbors who are unable or unwilling to protect their own interests, or to ensure that exemption permits will not be granted contrary to overall land use policies, including those for protecting solar access.

SECTION 4 Comprehensive Plan and Mapping

The [appropriate planning body] is directed to:

(a) Assess the likely rate, pattern, and extent of solar energy usage within the community as a whole and within its districts over a period of [25] years.

(b) Prepare a map showing the overlay zone types and the applicable heights and slopes of envelope planes for all areas of the City where overlay zones are to be applied, established in accordance with Section 5 of this Ordinance. A copy of this map, together with any amendments adopted from time to time, shall be provided to [all offices issuing development permits].

(c) Prepare an amendment to the City's Comprehensive Plan according to the procedures contained in [ordinance or statute setting out procedures to be followed in adopting the comprehensive plan]. The amendment shall encourage the use of solar energy in the City, shall incorporate the assessment and mapping developed under Subsections (a) and (b), and shall, to the extent feasible, be in harmony with other goals expressed in the Comprehensive Plan.

The [appropriate planning body] is directed to prepare the revision of the Comprehensive Plan and the first overlay zone mapping of the City within [six months] of the adoption of this Ordinance.

The [planning department] may prepare revisions to the overlay zone maps from time to time as refinements become feasible and desirable.

Plan revisions and overlay zone maps and changes shall become effective only when adopted by the [city council or whoever must approve zoning changes].

Comment on Comprehensive Plan and Mapping Clause. This prototype ordinance adopts what is at once the conservative position and the modern position: that a comprehensive plan independent of the zoning ordinance is needed to comply fully with zoning enabling legislation (see Chapter 3). By lending weight and logic to the divisions called for in Section 5 of this ordinance, a plan will help a community that uses solar access controls to protect itself against legal challenges, particularly those questioning whether the benefit the ordinance confers upon solar system owners is only incidental to the benefit to the public (see Chapter 3). If a locality has no comprehensive plan but wishes to adopt solar access protection similar to that described in this prototype legislation, it should still perform steps (a) and (c), but instead of (b) it should prepare a new comprehensive solar plan based on (a) and (c).

The most difficult step in applying this envelope approach is the mapping of a city. This will require a careful area-by-area examination of existing structures and vegetation and of structures that could be permitted under current controls. Mapping should take into account topography; height, bulk, and location of structures and vegetation; orientation of streets and structures; and other such factors. The material in Chapter 2 and Appendix B should be helpful in evaluating available access once basic data are collected.

Some rezoning where reasonable and politically feasible, in addition to adopting overlay zones, should be undertaken if the greatest possible protection is to be gained. Since an exemption of right is available where the underlying density cannot be achieved without breaching an envelope (see Section 7(b)), there is no point in adopting restrictions in overlay zones that are inconsistent with the densities of the underlying zone. (See the comment to Section 5 for further discussion of the mapping in terms of overlay zone types).

SECTION 5 Construction Method for Solar Access Envelopes

(a) In a Type 1 Overlay Zone, envelopes shall consist of a plane parallel to the base plane for the lot at the height specified for that overlay zone.

(b) In a Type 2 Overlay Zone, envelopes shall consist of a plane for each northern boundary line constructed as follows: the plane shall begin at a line parallel to the base line of the northern boundary line at the height specified for that overlay zone. The plane shall slope up inward over the property at the slope specified for that overlay zone; the slope of the plane shall be so measured as to be the least restrictive on development of either a slope (1) in relation to the horizontal or (2) in relation to the slope of the base plane in the direction perpendicular to the boundary line in question.

(c) In a Type 3 Overlay Zone, envelopes shall consist of the plane(s) for a Type 2 Zone plus a plane for each eastern and western boundary line constructed using the height and slope specified for that overlay zone for an eastern or western boundary plane.

(d) In a Type 4 Overlay Zone, envelopes shall consist of the plane(s) for a Type 3 Zone plus a plane for each southern boundary constructed using the height and slope specified for that overlay zone for a southern boundary plane.

Comment on construction method for solar access envelopes. While these envelopes are difficult to visualize from the words alone, the figures accompanying the introductory comment should make clear the features of envelopes in each type of overlay zone. It is not the intent of this report to suggest that all types of overlay zones have to be used. A single zone type could be used city-wide, with the height (and possibly the slope) varying with the conditions in the particular zone. The boundaries of these overlay zones could be the same or different from the existing land use zones, as appropriate.

A city would be making a useful step forward in solar access protection if it merely adopted Type 1 overlay zones for all areas of the city with planes at the maximum height for structures in each zone. Or a city could adopt overlay zones of Type 2, with sloping planes on northern boundary lines only, provided that the slopes and heights were chosen to be reasonable under the circumstances. Then, for example, if lots are fairly large, a fairly low slope and height which would allow early morning and late afternoon access on south walls (or even the ground below) could be adopted, allowing a new or expanded structure toward the center or south end of a lot so that the usual density for the neighborhood could be achieved. Provided that the height is not unreasonably low, a low angle could be used even in areas with fairly small lots. It is only necessary to adjust the height at which the sloping envelope plane starts so that the structure (and most vegetation) usual for the area will fit under the envelope.

Even the more complex overlay zone Types 3 and 4 could be adopted if the angles or the starting heights are not too low. Several cities have adopted or are considering bulk plane ordinances that are essentially of this type but that use steeply sloping planes (45 or 60 degrees). (See APA's *Guidebook for Planning Officials.*) An angle of zero degrees would presumably not be unreasonable at the present maximum permitted height for structures; in effect, that is what present zoning does. A low angle should be similarly reasonable if the envelope plane begins not too much lower than the current maximum permitted height.

The mapping should include both a map of the overlay zone boundaries and tables indicating applicable heights and slopes of envelope planes in the several zones. Table 5-1 illustrates how the table might look:

TABLE 5-1
APPLICABLE HEIGHT AND SLOPE OF ENVELOPE PLANES FOR VARIOUS OVERLAY ZONES

Overlay Zone†	Type	North Boundary Line Planes		East and West Boundary Line Planes	
		Height	Slope*	Height	Slope*
A1	1	80′		height overall	
A2	2	20′	3/7	NA‡	NA‡
A3	3	15′	3/7	10′	1/1
B3	3	25′	3/7	15′	1/1

* Slope is given in ratio of rise in plane height to distance from border line
† Ordinarily more zones would be used
‡ NA: not applicable

While more complex overlay zone maps could be worked out by hand, it may be better to await the development of computer programs that can do such mapping automatically and can accurately predict the overall extent of the resulting restraints on further development. This would allow more informed decisions that balance development with solar access. A complex initial mapping is not essential. This section allows for the planning department to propose new sets of overlay zones, heights, and slopes from time to time as appropriate.

If the most complex overlay zone types are to be adopted in the near future, they should generally be at conservatively high angles for the planes on the west and east (say, 45°) and south (say, 60°). These can be justified on the basis of providing a high standard of natural light and air, without regard to their bonus improvement in solar access protection. This would ensure that the overlay zoning would not be subject to criticism that it is premature and provides for remote future needs. Providing adequate natural light and air is a traditional public purpose for zoning.

SECTION 6 Effect and Enforcement of Solar Access Envelopes

(a) No City department shall issue any development permit that would allow the erection of any structure that would breach any solar access envelope over the lot in question, except to the extent that specific exemptions have been granted under Section 7 of this Ordinance. [Permit issuing departments] shall consult the map showing solar access overlay zones before issuing development permits. Any development permit that would result in a breaching of the applicable solar envelope is void except to the extent that it authorizes work that would not breach solar access envelopes or that is exempted under Section 7. A solar access beneficiary may protest or appeal the grant of a development permit that would purport to authorize a breaching structure. The protest or appeal may be made to [administrative agency or court of general jurisdiction].

(b) Pending the resolution of an appeal in court, a solar access beneficiary shall be entitled to a preliminary injunction to stop construction of such part of a structure as may breach the applicable envelope. The beneficiary shall not be required to post bond to obtain this injunction. The party having built or proposing to build the structure shall bear the burden of showing that the structure would not or does not breach the applicable solar access envelope or that an exemption under Section 7 applies. If the court finds that the structure would shade and breach and is not exempted, the injunction shall be made permanent and the development permit declared void to the extent that the structure would breach the envelope. If the building has already been built to the extent that it already breaches the solar access envelope, the court shall order removal of the breaching part if feasible or shall order the payment of damages to the beneficiary. The solar access beneficiary shall be entitled to the payment of damages where appropriate, court costs, and reasonable attorney's fees, as against either the City or the breaching party, or both.

(c) No one shall plant any vegetation that breaches a solar access envelope after an overlay zone is adopted except as exempted under Section 7. No one shall permit any vegetation within his control that [now is in place or] is planted after adoption of a solar access zone to grow in such a manner as to breach the applicable envelope except as exempted under Section 7. Property owners or registered lessees shall, not less than annually (or more often if necessary), remove or cause to have removed any part of vegetation that breaches a solar access envelope. [Add language on what pre-existing vegetation need not be trimmed—see comment for alternatives.]

(d) If vegetation is not trimmed as required in, or is planted contrary to, Subsection (c), a solar access beneficiary or the City on complaint by a beneficiary may give notice of the breach and shading by means of certified mail, return receipt requested, to the owner or registered lessee of the property where the breaching vegetation is located. If the property owner or lessee then fails to remove the breaching vegetation within 30 days after receiving this notice, an injunction shall be issued, upon complaint of the solar access beneficiary or the City. The injunction shall order the breaching party to trim the vegetation, and the court shall order the violating property owner or lessee to pay for any damages to the complainant, to pay court costs, and to pay the complainant's reasonable attorney's fees.

(e) If personal jurisdiction cannot be obtained over either the breaching property owner or registered lessee, the City may have a notice listing the property by owner, address, and [legal description], published once a week for four consecutive weeks in a newspaper of general circulation within the City, giving notice that vegetation located on the property is in violation of this Ordinance and is subject to mandatory trimming. The City shall then have the power, pursuant to a court order, to enter the property, trim or cause to have trimmed the breaching parts of the vegetation, and add the costs of the trimming, court costs, and other related costs due to the trimming to the tax liability attached to that property [OR: as a lien against that property].

(f) If the City is the property owner or registered lessee and has failed to trim breaching vegetation, a solar access beneficiary may notify the City [specific department] by certified mail, return receipt requested, asking the City to trim the vegetation. If the City fails to remove the breaching vegetation within

30 days after receiving this notice, the complaining beneficiary may bring suit for an injunction requiring trimming and may recover damages, court costs, and reasonable attorneys' fees from the City, if the City is found to be in violation.

(g) Where the property owner or registered lessee contends that particular vegetation is exempt from trimming requirements, the burden of proof shall be on the owner or lessee to show that an exemption applies to the particular vegetation.

(h) Notwithstanding any other provisions of this section, a solar access beneficiary may agree to allow any person to let vegetation grow in a manner that casts a shadow on the beneficiary's land. The agreement shall be in writing and signed by both parties. It shall be revocable at any time on not more than 90 days written notice, whether or not its terms so provide, and if the property owned by the solar access beneficiary is sold, the agreement becomes void. The City or another solar access beneficiary who has not made such an agreement with the breaching party may seek enforcement of this Ordinance as it affects the City or other beneficiary, notwithstanding such agreement.

(i) A property owner or registered lessee shall be entitled to reimbursement from the solar access beneficiary for all reasonable costs of keeping vegetation trimmed as required under this Ordinance, provided that there is no agreement between the parties under Subsection (h) and the landowner or lessee can show that the vegetation in question was in place on his property before the solar access overlay was adopted. [To be adopted where existing vegetation is to be grandfathered only to the extent of its existing growth. See Subsection (c)'s bracketed language.]

Comments on the effect and enforcement of solar access overlays. This and the following section are the heart of this prototype ordinance; its principal features have already been described in the introductory comment. The provisions for enforcement, especially as to vegetation, are lengthy but straightforward.

This prototype ordinance provides several specific remedies (injunctions, court-ordered entry and trimming by the city, damages, court costs, and attorney's fees). Some of these may be beyond the powers of a city to enact, depending on the state constitution and enabling legislation. The effectiveness of the overlay zoning depends in the first instance on careful administration by the city's permit-issuing agencies to protect against shadows from buildings. Without such remedies as provided here, however, such an ordinance might be useless. The protected parties need the power to seek enforcement when the city is unable or unwilling to enforce its regulations concerning structures; for vegetation, primary responsibility for enforcement should be on the citizen, not the city. Some cities may be able to adopt such an ordinance without further state enabling legislation. For many others, new enabling legislation would be needed. This report recommends state passage of legislation specifically authorizing such remedies to ensure that cities have the power to use them.

Since the police power has not traditionally been used to regulate privately owned vegetation in this manner, amendments to state enabling legislation may be necessary. The ordinance might have been written so that collector owners could enter their neighbor's land to trim the obstructing vegetation themselves. Easements sometimes allow similar privileges. ELI does not recommend self-help approaches as they could cause arguments over whether trimming was necessary and was done properly.

Another possibility is to declare shading vegetation a public or private nuisance, since a city or a private person may act to abate a nuisance (see Chapter 8). A city would also exercise its *in rem* jurisdiction (when *in personam* jurisdiction was not possible) to attach the property, and pay monetary damages from the proceeds; that seems far more drastic than is necessary.

Under Subsection (g), the owner of vegetation has the burden of proving that it was there before the solar access overlay zone was adopted. A simple way to do this would be with a notarized, dated photograph of the vegetation. When such an ordinance is enacted, the ordinance, or the regulations under it, could be written to advise people to prepare such proofs. Such proofs would also be useful under Subsection (i), which requires the collector owner to compensate neighbors for the cost of trimming in situations where the tree or bush predated the notice of the application for solar collector recordation.

Under Subsection (c) localities should choose the particular form of grandfathering of existing vegetation that is appropriate to local conditions:

- ☐ Existing vegetation that breaches the applicable envelope at the time of overlay zone adoption, but only to the extent of the size of the vegetation at that time (requiring trimming of further growth)
- ☐ Same as above but allowing for further growth without limit
- ☐ Existing vegetation, whether or not breaching the envelope at time of adoption, but only if the vegetation is larger than some minimum size (say, 3″ diameter trunk or 10′ tall) or more valuable than some minimum amount (say, $500) or
- ☐ Existing vegetation of whatever size (including seedlings), permitting all natural growth

ELI does not recommend that existing vegetation be removed or trimmed back to less than its height at the time of overlay zone adoption at this stage of solar energy use. Nor do we recommend the other extreme (the last of the options) of exempting all growth of any existing vegetation.

SECTION 7 Exemptions

(a) Such insubstantial breaches of a solar access envelope as utility poles, wires, flagpoles, and slender antennas are exempted. The [board of zoning appeals] may adopt regulations, or rule on individual cases, to define further insubstantial breaches by structures or vegetation that will be exempted.

(b) Any person who proposes to build a structure that would breach a solar access envelope may apply for an exemption permit under the procedure in Subsection (d) of this section. An applicant for an exemption permit shall be entitled to the permit as a matter of right if he can show that it is clearly infeasible to [achieve the underlying density provided in current zoning for the lot] without breaching the solar access overlay. Such exemption permit of right shall allow the least breach possible consistent with achievement of the density to which the applicant is entitled.

(c) Any person who proposes to plant vegetation or allow vegetation under his control to grow that would breach a solar access envelope may apply for an exemption permit under the following subsection.

(d) (1) Any property owner or lessee may apply for an exemption permit on an application of such form as the [Board] may prescribe; the application shall at a minimum include the following:

(A) The applicant's name and address, the owner's name and address, and the [local customary legal designation] of the property for which the exemption is sought.

(B) Survey plats or other accurate drawings showing property lines, boundary lines, dimensions, and topography of the property for which the exemption is sought and nearby properties that would be shaded by the proposed breaching structure and/or vegetation, as well as the locations and dimensions of all existing and proposed structures on these properties, and existing and proposed vegetation on the property for which exemption is sought.

(C) A certified copy of the official zoning map of [1,000] feet radius surrounding the property for which exemption is sought.

(D) A clear showing that the [underlying density under current zoning cannot be achieved] without breaching the solar access envelope and that the proposed structure minimizes the breach, if the application is for an exemption of right; otherwise, a clear showing that the nearby lots owned by solar access beneficiaries (excluding those for which protection is unavailable under Section 3(a)) are unsuitable for development that includes solar collectors because they are an historic district or parkland or because the existing vegetation, structures, topography, and other factors make the use of solar collectors impractical in any location that would be shaded by the proposed structures or vegetation.

(E) All other exhibits the applicant intends to introduce at the hearing.

(F) A summary of what the applicant expects to prove, including names of any witnesses, summaries of the testimony of any expert witnesses, and the estimated time to present the applicant's case to the Board.

(2) When an exemption permit application has been filed, the [board of zoning appeals] shall prepare a list of solar access beneficiaries who should be notified of the application because they may be affected by the exemption if granted. The [board] shall provide the applicant with a copy of this list.

(3) The permit applicant shall then certify to the [board] that the following actions have been taken, tendering proof satisfactory to the [board]:

(A) That a certified letter, return receipt requested, has been mailed to each solar access beneficiary on the list provided by the [board].

(B) That a sign has been posted on the property with the [board's] telephone number and "SOLAR ACCESS ENVELOPE EXCEPTION PENDING" in the form and for the period required under [section of zoning code setting out posting requirements for variance applications].

(C) The letter to solar access beneficiaries shall be on a form supplied by the [board] to make it clear what the applicant proposes, that the beneficiaries are entitled to protest the application, and what the effect of the exemption will be if granted; the [board's] telephone number to call for further information shall also be included.

(4) The [Board] shall set a hearing date and notify the applicant and any parties protesting the application.

(5) The hearing shall be held according to the rules of the [board], shall be on the record, and open to the public. The applicant for the exception permit shall bear the burden of proof on all issues. If the applicant makes the showing required by Subsection (d)(1)(D), the [board] shall (in the case of an exemption of right) or may (in the case of an other exemption) approve the exemption permit. The [board] may condition or modify any exemption as necessary to minimize the breach of the envelope consistent with the exemption to which the applicant is entitled.

(6) If the exemption permittee fails to make use of the exemption within a reasonable time, any solar access bene-

ficiary may apply to the [board] for a removal, in whole or in part, of any exemptions that are not of right or of any exemptions of right to which the permittee would no longer be entitled if the zoning has changed.

(7) Any losing party may appeal the determination of the [board] to [name of court of general jurisdiction in the city].

Comments on exemptions. The principal features of this section have already been described. Localities may have a "Board of Adjustments" or some other designation for the agency that hears variance applications (what here is termed the Board of Zoning Appeals). The procedures in this ordinance are largely based on typical procedures for a variance. Where local procedures for variances are different from those suggested here, the prototype might be adapted to the local procedures. It is recommended that the procedural safeguards found here (such as the notice by letter, posting, and advertising) be retained.

The bracketed phrase describing conditions required for an exemption of right in Subsection (b) could take a variety of forms other than the one here and still not constitute a taking (see Chapter 3). It could read (1) "to achieve the underlying density provided in current zoning for the lot;" (2) "to build a structure comparable in size to the average of structures already built in that zone;" (3) "to build a structure that provides some reasonable beneficial use." None of these phrases amounts to a clear standard, so amplification in an adopted ordinance or in regulations would be needed. If the type of standard in (2) is adopted, it may be necessary to allow for the constitutional minimum (3) where an area is not yet developed; ordinarily, (2) will be more liberal than (3) in developed areas.

[Further SECTIONS may provide severability, effective date, safety, and amendment clauses as desired.]

Comment on further sections. In addition to to the matters listed, if the local Planning Department or Board of Zoning Appeals does not have rulemaking power, it would be wise to provide for a power to make regulations to elaborate on the provisions of the ordinance and make their applications more intelligible to the average citizen.

6
Neighborhood-Scale Protection

The strategies discussed in this chapter will not provide the extensive protection available through zoning strategies. But they will usually affect more than one lot and can sometimes be used to guarantee solar access for large neighborhoods.

Two of these legal approaches—subdivision regulations and planned unit developments, will be of more use in new developments than in existing cities.[1] Impact statements should be useful in both situations. These strategies alone will not provide adequate solar access protection. But they could offer useful supplemental controls in cities where major redevelopment projects are common.

Subdivision Regulations

In many areas, changes in subdivision regulations offer a promising method to provide immediate solar access. It may be easier, politically, to change subdivision regulations than to change zoning. Modification of subdivision regulations for solar access protection is most effective in new, large developments. However, subdivision controls also have relevance to largely developed areas. The term "subdivision" may encompass both land division and land assemblage. Both occur in largely developed areas, although on a

1. *See* American Planning Association, Protecting Solar Access for Residential Developments: A Guidebook for Planning Officials (Washington, D.C.: U.S. Government Printing Office, 1979). Many suggestions for revising subdivision regulations to achieve better solar access protection may be found in this guidebook.

smaller scale than they do on virgin tracts. Furthermore, some villages and cities (e.g., Houston) that have no zoning do have subdivision regulations.

Definitions unclear. Subdivision regulations sometimes fail to define adequately what is meant by a subdivision and thus leave many questions for courts to decide. When a developer divides a parcel of land into smaller parcels, the project is usually subject to subdivision controls. However, laws differ as to how many smaller parcels must be formed and whether a new street must be formed before a "subdivision" occurs. When a developer aggregates small pieces of land into a larger assemblage for development the situation becomes murkier.

Subdivision controls versus zoning. Subdivision controls differ from zoning in several important ways. Zoning focuses on densities, bulk, and placement of buildings, and on the uses to which land may be put. Subdivision regulations are generally concerned with the manner in which a large parcel is subdivided into a number of smaller ones, including lot orientation, site design, street orientation and width, provision for parks, playgrounds, open spaces, storm drainage, and the like.[2] When parcels in largely developed areas are subdivided, it may be too late to make changes such as street orientation and width to benefit solar access. However, the provision of open space can still be usefully regulated. Open space requirements may be particularly useful where tall structures are being erected near shorter ones, unless the shorter ones are to the south. The location and sizes of open spaces should be made as specific as possible. If a large area is being redeveloped, it will sometimes be possible to require street plans and lot orientation that promote good solar access by requiring structures to have their longest axis in an east–west direction.

2. In fact, subdivision regulations sometimes serve many of the functions usually provided by zoning. Houston, one of the largest cities in the United States, has no zoning controls (i.e., has no ordinance that restricts the *uses* on various parcels). Bernard H. Siegan, *Non-zoning in Houston,* Journal of Law and Economics 13, no. 1 (April 1970): 75. Subdivision controls and restrictive covenants are used extensively, however. Nearly three-quarters of the buildings in Houston were erected subject to subdivision controls. Houston subdivision regulations provide, for example, that individually owned townhouses shall have setbacks of at least 20 feet and some screening. Single-family dwellings must have a setback of 25 feet, and must also meet some sideyard requirements. Ibid., p. 76. But because there are no rear yard requirements, some house backs are close together. Should Houston and other cities without zoning want to protect solar access, they could amend their relevant subdivision controls to assure appropriate street width and direction, structure orientation, and yard requirements.

Governments regulate subdivisions of land by requiring a developer to file a plan with the appropriate board, which must approve it before development begins. As a condition to approval, the community may require the developer to provide such amenities as streets, utilities, sewers, public land dedications, and particular lot patterns.[3] Under their existing statutory authority some communities could require, for example, the plans to contain easements protecting solar access, height restrictions, open space requirements, and controls on vegetation. Each locality should check its state statutes and court decisions, however, since new state enabling legislation may be needed.

Because different agencies typically administer subdivision and zoning controls, subdivision approvals are not always conditioned on compliance with zoning.[4] One authority states, however, that "it seems clear that plats should not be approved which violate existing zoning regulations."[5] Because there is uncertainty on this point, both subdivision laws and zoning laws should be amended to include solar access considerations.

Legislative possibilities. Several jurisdictions have legislation to allow or require subdivision plans that assure collectors access to direct sunlight. The Colorado Springs, Colorado, Planning Department tried a voluntary approach to solar access protection in the 1950s, without great success. Although developers who used solar collectors were offered setback waivers and flexibility in lot placement, as of February 1978, only one developer had taken advantage of this plan.[6]

3. *See* DONALD G. HAGMAN, URBAN PLANNING AND LAND DEVELOPMENT CONTROL LAW (St. Paul: West Publishing Co., 1971), §§ 138–139, pp. 253–59. In the past, such conditions have been imposed as a prerequisite to approval of the subdivision plan—a legitimate *quid pro quo* for the privilege of recording the developed parcels in the public land records. NORMAN WILLIAMS, JR., AMERICAN LAND PLANNING LAW, 5 vols. (Chicago: Callaghan & Co., 1975), 5: § 156.02, pp. 273–74. More recently, however, the privilege theory is being rejected and subdivision controls are justified as a legitimate use of the police power. HAGMAN, URBAN PLANNING AND LAND DEVELOPMENT CONTROL LAW, § 136, pp. 249–50.

4. *See*, e.g., Kane v. Zoning Bd. of Review, 97 R.I. 152, 196 A.2d 421 (1964).

5. ROBERT M. ANDERSON, AMERICAN LAW OF ZONING, 2d ed., 5 vols. (Rochester, N.Y.: Lawyers Co-operative Publishing Co., 1977), 4: § 23.21, p. 91.

6. PAS [Planning Advisory Service] MEMO, 1 April 1977, p. 3 (PAS MEMO is a publication of the American Society of Planning Officials, now the American Planning Association). The subdivision review provisions pertaining to solar access in the zoning ordinance of Colorado Springs read as follows:

> Solar orientation. If, for the reason of solar orientation, an entire area between two intersecting streets in a block is developed cooperatively or as a unit, all

In Connecticut, the statute that permits municipalities to regulate subdivisions was amended in early 1978 to add the following provision:

> Such regulations may also encourage energy-efficient patterns of development, the use of solar and other renewable forms of energy and energy conservation.[7]

The same language was also added to the statutes for municipal plans, zoning regulations, planned unit developments, and regional plans.

Proposed legislation in Minnesota would add the following statutory provision:

> A municipality may, for purposes of protecting and assuring access to direct sunlight for solar energy systems, prohibit or restrict development through subdivision regulations. Such regulations may call for subdivision development plans containing restrictive covenants, height restrictions, side yard and setback requirements, or other permissible forms of land use controls.[8]

It will be interesting to see if the nonmandatory approach taken in Connecticut and proposed in Minnesota works any better than the Colorado Springs experiment. If not, states should consider conditioning subdivision approvals on solar access protection. This is the thrust of a new California law that requires subdivision designs to provide for the use of future passive solar systems:

> The design of a subdivision for which a tentative map is required . . . shall provide, to the extent feasible, for future passive or natural heating or cooling opportunities in the subdivision.[9]

"Feasible" means capable of being accomplished in a reasonable time, taking economic, environmental, social and technological factors into account. Lot size and configuration to permit proper struc-

> yard regulations may be varied to carry out the purpose, providing that the zoning board of adjustment after public notice and hearing is of the opinion that such a development will not be injurious to adjacent property.

COLORADO SPRINGS, COLO., CODE § 15-41 (1951). *See also* the discussion of efforts in Colorado to require energy impact assessment in the subdivision approval process in KARIN HALVORSEN HILLHOUSE, LEGAL AND INSTITUTIONAL PERSPECTIVES ON SOLAR ENERGY IN COLORADO (Washington, D.C.: Environmental Law Institute, 1977), pp. 52–58.

7. Conn. Pub. Act No. 78-314, § 4 (1978) (to be codified as CONN. GEN. STAT. § 8-25).

8. Proposed amendment to MINN. STAT. § 462.359, subdiv. 2 (1976), in Minnesota Energy Agency, *Legislation Options for Encouraging Solar Energy Use In Minnesota: Executive Summary* (13 December 1977), p. 32 (available from Minnesota Energy Agency, 150 E. Kellogg Blvd., St. Paul, Minn. 55101).

9. California Solar Rights Act of 1978, 1978 Cal. Stats. ch. 1154, § 8 (to be codified as CAL. GOV'T CODE § 66473.1).

ture orientation and consideration of local climate and land contours are given as examples of design opportunities. The law states, however, that such provisions shall not result in reducing allowable densities or the percentage of a lot which may be occupied by a building. The legislative bodies of California cities and counties are ordered to deny approval of final or tentative maps if they find that the design of a proposed subdivision does not, to the maximum extent possible, provide for the future use of passive solar systems.

Conclusions. Subdivision controls to provide solar access in largely developed areas would affect only a few units. Whether the approval of subdivision plans can be conditioned on the provision of amenities, such as solar access, depends on the language of the state statute authorizing subdivision regulations. Yet another drawback is that once a subdivision development is complete and sold, the subdivision laws no longer apply to it. Thus, unless restrictive covenants provided otherwise, the new owners could plant trees anywhere they liked or could make whatever structural additions they wished as long as they complied with other ordinances.

On the other hand, subdivision regulations to protect solar access are likely to be politically acceptable and are certainly compatible with other public policies, such as the promotion of energy conservation. The implementation of subdivision regulations is relatively simple, the costs involved could be passed on to purchasers rather than to the city, and regulations could be written to give clear notice of rights and duties. If conflicts occurred, administrative remedies could be available to make court proceedings unnecessary. Although none of the state legislation described above specifically mentions vegetation, language could be added to such bills to make it clear that subdivision ordinances may require developers to protect solar access when planting new vegetation.

Impact Statements

The federal government and some states and localities require an analysis in an impact statement of the probable environmental consequences of certain activities before those activities may be undertaken. The impact statement process could be relevant to solar access because: (1) the effect of a project on available solar access is an environmental impact that arguably should be considered under existing laws; and (2) the procedures used to assure consideration of environmental factors could be adapted for use in separate solar access laws (by requiring a "solar access impact statement" to be prepared before any significant development—private or public—could be undertaken).

The prime and original example of a law requiring environmental impact statements (EISs) is the National Environmental Policy Act (NEPA).[10] It requires EISs to be prepared for every recommendation, report, or proposal for legislation and other major federal actions that would significantly affect the quality of the human environment.[11] (NEPA applies only to federal actions not to private endeavors. Some of the laws modeled on NEPA, however, do apply to private actions.)

One factor that must be specifically addressed under NEPA is any irreversible commitment of resources that would result from implementing the proposed action.[12] The source of energy to power a new development or structure involves a commitment of resources that is difficult to reverse. In an environmental impact statement for construction of a new federal structure, therefore, alternative heating and cooling systems merit consideration. Arguably, if the federal structure blocked an existing collector's access to solar energy, this environmental impact must also be noted in the EIS, and alternative building configurations to mitigate this impact considered.[13]

About twenty-five states also have laws or regulations modeled on the federal EIS procedure.[14] Nine of these require an analysis of the energy demand that would be imposed by the suggested project and of sources of energy to meet that demand.[15] State attempts to re-

10. 42 U.S.C. § 4321 et seq. (1970). For reviews of experiences with NEPA, *see* FREDERICK R. ANDERSON, NEPA IN THE COURTS (Baltimore: Johns Hopkins Press, 1973); U.S. COUNCIL ON ENVIRONMENTAL QUALITY, ENVIRONMENTAL IMPACT STATEMENTS: AN ANALYSIS OF SIX YEARS' EXPERIENCE BY SEVENTY FEDERAL AGENCIES (Washington, D.C., 1976); and RICHARD A. LIROFF, A NATIONAL POLICY FOR THE ENVIRONMENT: NEPA AND ITS AFTERMATH (Bloomington, Ind.: Indiana University Press, 1976).

11. Ibid., § 4332(2)(c).

12. Ibid.

13. By itself, the blocking of access might be only a minor environmental impact and would probably not require the preparation of an EIS. Presumably, however, for any new federal building an EIS would be needed because of the building's cumulative environmental impacts.

14. *See,* e.g., Nicholas C. Yost, *NEPA's Progeny: State Environmental Policy Acts,* ENVIR. L. REP. 3 (1973): 50090–98; and Ralph Catalano and Joseph DiMento, *Local Government Response to State Environmental Impact Assessment Requirements: An Explanation and a Typology,* ENVIRONMENTAL LAW 7 (1976): 25–50.

15. *See* Michael Gerrard, *Disclosure of Hidden Energy Demands: A New Challenge for NEPA,* ENVIRONMENTAL AFFAIRS 4 (1974): 665–66. Somewhat ironically, the *removal* of impact statement requirements can also be used to spur solar development. California A.B. 3476 (1978) would exempt any project for construction and operation of certain types of solar energy power facilities from having to prepare impact reports.

quire impact statements have not been uniformly satisfactory.[16] To lessen the burden on developers somewhat, consideration of solar access impact could be folded into the building permit process. (When a building permit is sought in California, for example, an environmental impact report form is prepared at the same time.)

Some towns and counties have enacted NEPA-type legislation, either under powers granted them by special state enabling legislation, home rule charter, or under their authority to control zoning and subdivisions and to review site plans.[17]

To be a useful aid to persons making decisions on solar access, impact procedures should be integrated with comprehensive planning and other land use controls. Factors that localities typically consider during an EIS process include many considerations relevant to comprehensive plans: probable impacts of the proposed activity on the human and natural environments, alternatives to the proposed action, and tradeoffs between short-term versus long-term benefits.

The EIS process can also be adapted to include specific consideration of energy issues. In California, for example, an environmental impact report must include a detailed description of any "[m]itigation measures proposed to minimize the significant environmental effects including, but not limited to, measures to reduce wasteful, inefficient, and unnecessary consumption of energy."[18] Unlike most other states and unlike NEPA, California imposes substantive as well as procedural obligations: state agencies should not approve proposed projects if there are feasible alternatives or mitigation measures available that would substantially lessen the project's environmental effects.[19]

16. *See*, e.g., Donald G. Hagman, *NEPA's Progeny Inhabit the States: Were the Genes Defective?* Urban Law Annual 7 (1974): 3-56. In New Mexico, for example, a statute modeled closely after NEPA was repealed due to pressure brought by industrial and state agency representatives. *See* Charles R. McCash, *The Rise and Demise of the New Mexico Environmental Quality Act, 'Little NEPA,'* Natural Resources Journal 14 (1974): 401-09.

17. Examples are Troy, Michigan; Clearwater, Florida; Bedminster, New Jersey; Chatam, New Jersey; Monmouth Beach, New Jersey; Forsyth County, North Carolina; and Westchester County, New York. *See* Richard Oliver Brooks, Municipal Environmental Ordinances, 2 vols. (Kingston, R.I.: University of Rhode Island, 1978), 2: 365-97, for a discussion of local environmental impact strategies.

18. Cal. Pub. Res. Code § 21100(c), as discussed in Corbin Crews Harwood, Using Land to Save Energy (Cambridge, Mass.: Ballinger Publishing Co., 1977), pp. 195, 199-200.

19. Cal. Pub. Res. Code § 21002 (1976). *But see* Laurel Hills Homeowners' Ass'n v. City Council of Los Angeles, 147 Cal. Rptr. 842, 8 Envir. L. Rep. 20714 (Ct.

Perhaps equally significant is the provision in the California law that extends the environmental impact report procedure to activities on private land, as well as to zoning ordinances, zoning variances, conditional use permits, and approval of tentative subdivision maps.[20] Interestingly, most local EIS statement requirements apply to private, not public, projects.[21]

An aggressive state government concerned about long-term land uses that are compatible with the installation and use of solar energy systems could enact legislation requiring solar-energy-use impact statements. Although such legislation is legally possible, there is likely to be political opposition to imposing yet another permit requirement. Before a locality decides to enact such a law, it should carefully study enabling laws and its home-rule charter, if any, to determine if it has the requisite authority.

Application to vegetation. Existing impact statement procedures are clearly applicable to structures. This process has not, however, been applied to individual trees on public or private property. Yet it is precisely such trees that will cause most solar access problems in low-rise development. Impact statements for each proposed tree are not a feasible way to regulate vegetation shadows because of the administrative burdens of processing them. Such requirements would not be likely to attract the support of legislators. On the other hand, if a city is planning to plant rows of trees along parking strips or other public lands in developed areas, it could readily analyze the solar access impact of the proposed plantings.

Conclusions. States and localities that already have impact statement machinery should amend those laws and regulations to include the consideration of impacts on solar access. The additional work required would be small and the benefits would justify the expense. However, even states that have enacted EIS laws typically require only procedural, not substantive, compliance; thus, there would usually be no assurance of access to sunshine. Furthermore, states have evidenced a reluctance to apply impact statement requirements to private as well as public projects (although the local trend is the reverse). Most solar access disputes will involve private property.

App. 29 Dist. 1978), where it was held that this provision does not prevent the approval of a subdivision plan if mitigation measures will reduce the environmental effects to an "acceptable" level (despite the existence of environmentally superior alternatives). The California Supreme Court unanimously denied a petition to review the court of appeals decision, 27 September 1978.

20. Cal. Pub. Res. Code § 21080(a).

21. Brooks, Municipal Environmental Ordinances, 2: 368.

ELI does not recommend separate, new, "solar impact statements":

- □ Unless carefully integrated with other land use controls, such as comprehensive planning, solar access impact statements could result in case-by-case decisions that are vulnerable to court challenges on constitutional grounds. Without this precaution, they could also result in irrational land development patterns.
- □ Environmental impact statements are most helpful in generating information and identifying fairly general possible impacts of proposed projects. But the success of solar access decisions may hinge on a few degrees of exposure. A more precise tool seems better suited to this task.
- □ Only about half of the states have enacted NEPA-type laws. This may suggest that there has been political resistance to such laws (or it could reflect an apathy toward environmental concerns).
- □ The focus of impact statements is on how proposed projects will affect surrounding areas. Yet in solar access provision, the critical issue is how neighbors will affect the proposed collectors. At the time collectors were installed, their owner would have no way of knowing whether access to sunshine was secure.
- □ Impact statements weigh many factors against one another, rather than flatly prohibit any particular consequences. This flexibility introduces uncertainty that would discourage investment in solar energy systems.

Planned Unit Developments

Cities that anticipate significant redevelopment within their limits should consider using the flexible zoning technique of a planned unit development (PUD) to provide for solar access. PUDs offer an excellent tool in this regard because of their emphasis on a single, overall plan for a project, and because of their relaxation of building design, setback, height, yard, density requirements, and lot size. (PUDs can be any size.)

While the term PUD describes a type of development, it also applies to a regulatory process. Government administrators use site plan reviews, bonuses, and special permit, overlay zone, or floating zone controls to bargain for the elements incorporated in a PUD.

Unlike traditional zoning or subdivision controls, where land use decisions are often made far in advance of actual development, PUD decisions are made only after a development is proposed. Another distinction is that administrators are given much more discretion with regard to PUD approvals.

Although PUDs are implemented largely by local governments, state enabling legislation is advisable.[22] The enabling legislation should specify solar access provision as a goal. Standards in a PUD ordinance should state the amount of solar access which must be available, as well as other layout, design, and landscaping requirements. Performance, rather than prescriptive, standards will allow a developer maximum flexibility. Other possibly relevant local laws—such as those controlling special permits, overlay zones, or floating zones—should be made consistent with the provision of solar access. Because many developers may not understand the technical points of solar access, a locality could require preapplication conferences between planners and developers. Holding such conferences will help avoid costly and time-consuming changes to formal plans.

Solar access provided through the PUD site-design process could be obliterated if purchasers of units or structures in the development planted vegetation where it could shade collectors. A possible way around this problem is to have the PUD ordinance require private covenants to control the size and placement of vegetation. Proposed covenants could be scrutinized as part of the site plan review.

APA's *Guidebook for Planning Officials* suggests language to modify local PUD ordinances to make them suitable for solar access protection. Their *Manual for Developers* offers solar design standards useful to architects and developers who are laying out PUDs. Another method for figuring available solar access is suggested in Appendix A. Excellent discussions of the general pros and cons of PUDs and of their legality and administration are readily available.[23]

22. Some type of PUD enabling legislation exists in at least eighteen states. However, many cities use the PUD device even in the absence of enabling legislation. FRANK S. SO ET AL., PLANNED UNIT DEVELOPMENT ORDINANCES, Planning Advisory Service Report no. 291 (Chicago: American Society of Planning Officials, 1973), p. 47.

23. *See, e.g.*, MICHAEL J. MESHENBERG, THE ADMINISTRATION OF FLEXIBLE ZONING TECHNIQUES, Planning Advisory Service Report no. 318 (Chicago: American Society of Planning Officials, 1976); ANDERSON, AMERICAN LAW OF ZONING, 2: §§ 11.09–11.24, pp. 358–91; and HARWOOD, USING LAND TO SAVE ENERGY, pp. 107–16.

Part III
Lot-By-Lot Protection

The strategies described in the following three chapters share an attribute that is often seen as desirable: they will protect solar access only when and where it is immediately needed on a particular lot. A local government could allow development to proceed as usual in all other areas. Thus, at least initially, less of a burden is imposed on both the government and private developers, and lower public expenditures are required.

Another reason these strategies merit consideration, is that situations will occur when it is simply impossible to make generalizations about the physical characteristics of lots in an area—in other words, no (or very few) lots seem to be "similarly situated." Solar access zones, which treat similar lots alike, do not make sense in such areas. If private remedies are insufficient, then some type of public law is needed.

Lot-by-lot solar access protection strategies do not merit acclamation, however:

- □ They do nothing to protect solar access that is likely to be needed in the future.
- □ They require much individual effort: to secure a right to solar energy people must actively claim it, either by filing complicated forms or by suing or prompting the government to sue on their behalf. Under the approaches in Part II, on the other hand, landowners do not have to take the initiative; their rights and duties are predetermined.

- It is difficult, if not impossible, to integrate these strategies with rational land use planning.
- The first-person-to-develop-wins nature of the approaches may lead to challenges on constitutional grounds. The owners of identical lots may have very different restrictions placed on their lands solely because their neighbors take different actions.
- Since development rights can be drastically affected by the development of a neighbor's lot in relation to the development of one's own lot, interesting questions are also raised as to the duties of someone such as a tenant or trustee who has possession of land (but does not own it) to act to protect the owner's development or solar access rights.
- Because only a few property owners (those who install collectors) might benefit in any particular area, lawsuits claiming lack of a valid purpose may occur.

Careful drafting of a lot-by-lot protection law could sometimes reduce the hardship on the burdened neighbor of a solar collector. For example, the amount of solar access protection which would be granted could be specified in the law, even though it is not known which particular lots would benefit. A well-written law would not just say "shadows on collectors are public nuisances;" it would specify what types of collectors merit protection, and what areas were to be shadow-free during what hours. At a minimum, a law should consider factors such as the slope of land and the actual and allowed densities of land. In this manner, a law could help ensure that similar restrictions and benefits adhere to similarly situated lots and would avoid lawsuits claiming a violation of equal protection or due process requirements.

We are indebted to William A. Thomas of the American Bar Foundation for the following hypothetical situation about what could occur in the absence of both careful drafting and good will between neighbors:

Mr. Bury purchased an undeveloped lot immediately north of the undeveloped lot belonging to Mr. Pope. He then constructed a magnificent edifice with an extended east-west orientation just north of the property line and installed a solar energy collector at ground level at each end of the house. The local solar access law declared all shadows on collectors anywhere to be illegal. Mr. Bury's collectors therefore effectively precluded Mr. Pope, his neighbor, from developing his property while affording Bury an unobstructed view of the landscape over Pope's property as well as unobstructed solar skyspace. What are Pope's options following what, prior to access law, would have been considered "Bury's Folly?"

- Retain the undeveloped property at its greatly reduced value.
- Sell the property and purchase unencumbered property elsewhere.
- Pay Bury to remove the solar energy collectors.

Bury's options, on the other hand, are much more favorable:

- Continue to reap the free benefit conferred by the expensive burden placed on Pope.
- Purchase Pope's lot at its greatly reduced value and perhaps remove the solar energy collector and resell the lot at a handsome profit.
- Hold out for an exorbitant private settlement to remove the collector or waive the right, thus forcing Pope to reacquire a property right lost without any volition, illegal activity, or recompense.

It is unlikely that Bury would be willing to sell the new property right for what might be a reasonable amount in the present market, because both the advantages of having a solar energy collector and the value of the new right will greatly increase in the future. Few laws are so aptly designed to promote acrimonious relations between neighbors.

When considering the lot-by-lot protection strategies of collector recordation, nuisance law, prescriptive easements, "instant access" rights, and water law, keep in mind the problems raised by the above hypothetical, and whether good drafting alone could solve them. Although ELI does not really recommend any of these strategies, we favor collector recordation over the others. Even this approach, however, cannot effectively protect access that will be needed in the future, or remove all uncertainties as to constitutionality.

7
Permits and Recordations

If only lot-by-lot protection is wanted, the problem is to find the best way to use the police power to allow solar access to be "staked out" for the benefit of individual lots and made part of the public record. Several types of legal controls are possible vehicles: building permits, special permits, floating zones, and a new hybrid device called "recordation." (The prototype legislation at the end of this chapter uses the last device. It is termed recordation because the existence of a collector is recorded in the land records.)

Under any of these approaches, the size and shape of the protected area could be defined however the local government wished: the area to be protected could depend on the location of the collector needing solar access, or it could be a premeasured amount of solar access which would apply to all lots subject to access controls. It could even take the shape of solar envelope controls, which would be applied to adjacent properties. Although the protected areas under these approaches could be defined in the same way that they are for zoning-based approaches, the difference is that controls would not be applied until they were needed to protect a specific actual or proposed collector.

Building Permits

The primary function of building permit requirements is to secure compliance with building codes by allowing only safe structures to be built. Courts upheld such safety purposes of building codes as valid uses of the police power even before zoning existed.[1] Another important purpose of building permits is to ensure compliance with zoning codes.[2]

Building permits may often be required before structures are fitted with solar collectors.[3] The fact that a permit is already required suggests the possibility of providing solar access permits as part of the same process. Efficiency and economy can both be improved if permitting requirements can be combined.[4]

An attraction of using building permits to secure solar access is that local citizens may be more amenable to changes in building codes than they sometimes are to proposed zoning changes. Building permits also convey somewhat more of a right against the government than does zoning. Zoning laws do not give citizens the right to keep a particular zoning of their land. Some people fear, therefore, that solar access laws based on zoning would not give potential collector owners a secure feeling about investing in solar equipment. A building permit, however, can give its holder a limited type of property right. A right to complete a building arises under an estoppel theory if expenditures are made on actual construction after the permit has been issued.[5] Even if zoning is changed, the building may be completed (and may remain) as a nonconforming use.

1. People ex rel. Namm v. Carlin, 182 App. Div 626, 169 N.Y.S. 295 (1918).

2. An example of a case approving this procedure is Bon Air Estates v. Village of Suffern, 32 App. Div. 2d 921, 302 N.Y.S. 2d 304 (1969).

3. *See generally* FRANK O. MEEKER III, BUILDING CODES AS BARRIERS TO SOLAR HEATING AND COOLING OF BUILDINGS (Washington, D.C.: Environmental Law Institute, 1978). Building permits are usually needed when a building is constructed or an existing building is repaired, remodelled, or added to. *See,* e.g., State v. Walker, 265 N.C. 482, 144 S.E.2d 419 (1965); Farmers and Planters' Co. v. Mayor and Council of Salisbury, 136 Md. 617, 111 A.112 (Ct. App. 1920); Lazich v. Butte, 116 Mont. 386, 154 P.2d 260 (1944). Changes like the installation of some air conditioning systems, General Bowling Corp. v. Mount Lebanon, 373 A.2d 783 (Pa. Commw. Ct. 1977), or the addition of accessory structures are also regulated by most building codes.

4. *See generally* FRED BOSSELMAN ET AL., THE PERMIT EXPLOSION: COORDINATION OF THE PROLIFERATION (Washington, D.C.: Urban Land Institute, 1976). That study recommends coordinating state permitting programs by using a synthesis of the state of Washington's coordinated permit procedure, the American Land Institute's Model Development Code, and other specified systems. On the local level, the authors conclude that "while there is a trend toward the use of hearing officers to obtain facts on which to base such decisions [regarding individual permits], efforts to coordinate all permitting procedures with actions of local legislative bodies does create additional problems." Ibid., p. 3. Dade County, Florida, is discussed as an example of how local permit procedures may be consolidated.

5. However, the mere application for a building permit does not perfect a vested right. For example, if the zoning law is changed before a permit is issued and sub-

It can be argued that solar access permits could provide an analogous right to continue to receive access protection even in the event of a zoning change. Once there has been substantial reliance on a solar access permit, courts may find that a property right exists and therefore may require compensation if that right is taken away by the city. (More likely, courts would find the change in the law invalid as applied to the property of the permit holder.[6]) The weakness in this analysis is that solar access permits would restrict adjacent properties: they would convey a right against private parties as well as against the government. It is unknown whether courts would allow the restrictions against neighboring lands to remain as "nonconforming restrictions" (a somewhat novel idea) in the event of a rezoning.

stantial work performed in reliance on the zoning alone, courts may find that no right to construct exists. An example is a recent California Supreme Court Case, Arco Community Developers v. South Coast Regional Comm'n, 17 Cal. 3d 577 (1976).

6. *See* Gruber v. Township of Raritan, 39 N.J. 1, 186 A.2d 489 (1962).

7. Mass. S.B. 296 (1977). (More recent efforts in Massachusetts no longer rely on the building code; the board of zoning appeals would assume largely the same responsibilities outlined here. *See* Mass. S.B. 304, H.B. 5711 (1978).) Under this proposed legislation, the building permit process is used as the central means of guaranteeing solar access. The language providing solar access is made part of the state building code itself. The bill outlines two sets of procedures, one for areas zoned for four or fewer stories. All property owners who want to install collectors would first have to get a building permit under a special procedure. Once a permit is secured, no other development permits will be issued if they would interfere with solar access for the permitted collector. But a permit cannot be issued unless a solar energy unit is "sited and designed so as to be at a reasonable height, location, and angle to the sun in order to minimize the possibility that future development of nearby property will result in interference with the solar energy unit's access to solar energy" (§ 29). Although Section 30 puts the burden on an opposing party to show that the requirements of Section 29 are not met, the bill offers no guidance as to what is reasonable. This would pose a significant problem for administrators trying to implement such legislation. In areas of four or fewer stories, a building permit for the solar unit must be issued unless it can be shown that such requirements as collector size and location are not met or that a neighbor's "substantial reliance interest" would be destroyed if the permit were granted. Neighboring property owners are sent written notice of a building permit application only if future activities on the neighbor's land could be expected to interfere with the proposed collector's access to sunlight. Neighbors are given an opportunity to object but only on this ground. The effect of these provisions makes it easy to obtain sunlight protection in low-density areas. Speculative injury to property value does not constitute a substantial reliance interest. In areas zoned for more than four stories, building permit applications are referred to the municipal planning agency. The applicant gives notice to all landowners against whose future development protection is sought. The planning agency holds hearings and makes a recommendation on whether to approve the permit. The municipal governing body can either approve this recommendation or overrule it. (As applied to areas zoned for more than four stories, the proposal is not a true building code device.)

Perhaps because of the attractive qualities of building codes, at least two states have considered their use for solar access protection.[7] In spite of the attributes of building codes that make them seem appropriate vehicles for solar access protection, ELI does not believe they should be used for this purpose:

- □ Building codes do not regulate vegetation. Permits are only required when structures are built. A provision requiring people to obtain a building permit each time they want to plant a tree or bush is not in the public interest. It would discourage plantings, be expensive, and be difficult or impossible to enforce and administer.
- □ Building codes, which emphasize safety, are not the logical vehicle for solar access provisions. Access laws will have a major impact on land use patterns and should therefore be integrated with the laws designed to control or influence such patterns. If they are not, they may work at cross-purposes with these laws.
- □ Building code solutions would be afflicted by the difficulties that attend any lot-by-lot approach: the need to define solar collectors, the vulnerability to judicial questions about due process and equal protection, and the failure to protect future solar access.

Special Permits

Nearly all zoning enabling acts and ordinances authorize cities to use special permits.[8] The Standard State Zoning Enabling Act provides, in Section 7, the power to "hear and decide special exceptions to the terms of the ordinance upon which such board is required to pass under such ordinance." Special permits are the means usually employed to exercise this power.[9]

Special permits may be required before particular uses can be established in an area. While these uses are neither absolutely permitted nor forbidden in the zone, they have special qualities that require sensitivity in site selection because of their impact upon nearby

8. Donald G. Hagman, Urban Planning and Land Development Control Law (St. Paul: West Publishing Co., 1971), § 114, p. 208.

9. Sometimes the term "special exception" is used instead of special permit, and sometimes no special name at all is given to the required approval. Robert M. Anderson, American Law of Zoning, 2d ed., 5 vols (Rochester, N.Y.: Lawyers Co-operative Publishing Co., 1976), § 19.01, p. 358.

parcels. Typical examples of uses for which special permits may be required are public utility substations, gasoline stations, schools, churches, parking lots, airports, and recreation facilities. Solar collectors will also have an impact on surrounding properties: unlike the usual special permits, however, solar access special permits would actually proscribe activity on adjacent sites. A few instances do exist where special permits are used to protect the permitted use from activities on adjacent land. An example is a California ordinance requiring special permits before homes may be built in industrial areas.[10] This is also a poor analogy, however: residences in industrial areas require permits because they are out of place and impose burdens on nearby lots. Solar collectors will not be out of place. To treat them as such might pose a psychological barrier to their use and be used as evidence against collector owners in lawsuits attempting to ban collectors on aesthetic grounds.[11]

An earlier draft of the prototype legislation at the end of this chapter used special permits as the basic mechanism for solar access provision because the underlying purpose—to regulate uses requiring sensitivity in site selection—had a surface compatibility with solar access provision. The familiarity of local governments with special permits also made them seem an attractive choice. Furthermore, the conditions that must be met to obtain a special permit do not change the zoning of the property, so special permits are not challenged as spot zoning.[12]

However, to adapt the special permit process to protect solar access would fundamentally change its nature. Trying to attach a familiar label to what is really a new legal control generates confusion. Therefore, ELI believes that this discussion of special permits is useful chiefly as an example of the great flexibility of police power controls and as an example of how this power can legitimately be applied to allow uses in the public interest that may reduce the value of surrounding properties.

10. In Roney v. Board of Supervisors, 138 Cal. App. 2d 740, 292 P.2d 529 (1956), a California court found an ordinance valid that required special permits before homes could be put in heavy industrial areas. Permits would be refused if an area was too dangerous.

11. There may be heavy industrial areas where the air is so dirty that collector surfaces could not economically be kept clean. In such areas they might, indeed, be out of place.

12. *See* Fitzgerald v. Board of Review of Newport, 206 A.2d 635 (R.I. 1965); Bruzzi v. Board of Appeals, 84 R.I. 220, 122 A.2d 877 (1956).

Floating Zones

Floating zones are like other zones, except they are usually smaller in size and are not affixed to specific land until the landowner's application for that zoning is approved. These zones are most frequently used to allow shopping centers and planned unit developments, although they have also been applied to the conservation of land for parks and recreation purposes. Any landowner whose property meets a given set of conditions may apply. For a floating zone to be approved, it must meet the ordinances' standards. Site plan reviews are generally required, as is an ordinance amendment to attach the zone to particular land. Several of the variables typically regulated by floating zone ordinances are relevant to solar access provision: height limits, landscaping, and maximum lot coverage.[13]

As is the case with special permits, there is a superficial compatibility between floating zones and the requirements of a good access law. The flexibility of floating zones (combined with clear standards) is attractive. Such zones would seem to offer more safeguards against discrimination than would conventional lot rezonings. The basic notion of having a zone "float" until it is actually needed somewhere also seems appropriate to solar access. But "solar access floating zones" would differ from conventional ones in an important way: when they became anchored to particular land they would have to apply not only to the property of the applicant but also to neighboring lots. All the benefit would go to the applicant; all the detriment to the applicant's neighbors.

Another basic problem is that if floating zones are extensively used, they may be seen as promoting *ad hoc* land use decisions and violating comprehensive planning.[14] From a legal viewpoint this is somewhat ironic: the greater the use of solar energy, the less justifiable this type of access law would become. (Normally, the greater the number of persons using solar equipment, the greater the public purpose—and hence validity—of a police power control.)

Floating zones are not recommended as a way to provide solar access protection. But communities with floating zone ordinances should amend those laws and accompanying standards to include

13. *See* Michael J. Meshenberg, The Administration of Flexible Zoning Techniques, Planning Advisory Service report no. 318 (Chicago: American Society of Planning Officials, 1976), p. 30 ff.

14. Ibid., p. 32.

solar access concerns, so that the zones will not be approved if the resulting construction would shade collectors, and so that the new structures within the zone will be designed and oriented with solar access in mind.

Solar Collector Recordations

The procedure for protecting solar access described in the prototype legislation is original to this study. Its major distinguishing characteristics are as follows:

- The existence of solar collectors may be recorded in public records.
- As a result of that recordation, a specified, reasonable amount of solar access will be protected.
- To provide access, the development potential of some lots near a recorded collector will be reduced by a small amount.
- No monetary compensation will be paid the owners of these lots.

When trying to protect solar access on a lot-by-lot basis, the hardest problem to deal with, from a legal perspective, is that the government is seeking to take away development rights from N (the neighbor of the collector owner) to benefit O (the owner). As one reviewer notes, this fundamentally changes property law:

> O cannot simply claim negative easement rights unless N grants them to O. On what basis can you give O rights he doesn't have today? If you do not compensate N, you must answer both that the pursuit of energy conservation goals is a proper police power activity (I agree) and that the diminution of N's interest is akin to that suffered in conventional zoning situations (not as clear).[15]

The latter point is troublesome, but surmountable. By carefully limiting access protection to only reasonable protection, the burden on N will be minimal. In fact, in many cases the restraint will be more apparent than real. The prototype legislation carefully limits the amount of solar access that is protected, so that everyone can build structures to the approximate densities enjoyed by their neighbors. Existing structures and vegetation are exempt. The police power can clearly impose the minor restrictions envisioned in the prototype legislation on property when there is a public purpose for

15. John J. Costonis, in his written review of a draft of this report.

doing so. A separate question is whether this can be done on a lot-by-lot basis. ELI believes that it probably can, although there is no case law yet to support this belief. Such uses of the police power as special permits and floating zones have been upheld even though they often result in a greater depreciation of property values in areas surrounding the parcel to which they apply than solar access is likely to. Admittedly there is a distinction: solar access laws would restrict the ways nearby parcels could be developed. But the chief consequence of those restrictions is a reduction in the market value of the land—the same consequence as the granting of an application for a special permit or floating zone.

As the prototype legislation uses the police power in a slightly new way, there is a small risk that courts will find the use invalid. Governments worried about this possibility may wish to take one of two steps: (1) base their access legislation on their common law powers as well as their police power; or (2) provide nonmonetary compensation.

Common law backup. Before there were zoning or other land use controls, aggrieved property owners had to rely on nuisance suits. Nuisance law is an imperfect solar access protection device but it could be serviceable if carefully controlled. A locality could add a clause to its solar access ordinance declaring the shadows prohibited (through the police power) by a collector recordation law to also be either public or private nuisances. Such legislative findings that something is a nuisance will usually be upheld by courts. (Chapter 8 discusses the pros and cons of nuisance law in more detail.)

Provision of nonmonetary compensation. Compensatory approaches are covered in Part IV, but a limited type of "compensation" deserves mention here. It is less likely to be said that a government is making a gift of public funds to an individual if no money is paid. One possibility is to give a property owner development rights to replace those taken away. Transferable development rights (TDRs) schemes are one option (see Chapter 11). A more modest option that we call mini-TDRs is best explained by a hypothetical example: Trepidation City decides to protect solar access as needed, using a customized version of the prototype legislation. The city planners and attorneys are concerned that the unevenness of the burdens imposed may stimulate challenges. To reduce or eliminate the burden on persons who own land near collectors, it is decided that the city will give as much as it takes. In other words, if development of Blackacre Estate must be limited on the north to allow Sunnyacre's collectors

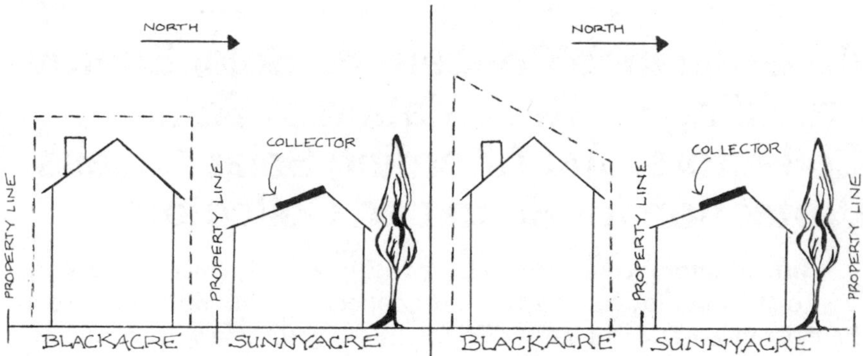

Development Originally Permitted (within dotted lines)

Development Permitted Under New Solar Access Law (within dotted lines)

FIGURE 7-1
BLACKACRE ESTATE EXAMPLE

to function, then Blackacre may be developed more intensively on its southern parts than formerly permitted (see Figure 7-1). The volume of area within which development may occur remains the same—only its location changes. Note that the structure on Blackacre was not as large as allowed under the old zoning (if it were, then no structural development rights would be lost when the new law went into effect). Although the change in development rights looks dramatic on paper, the actual repercussions will be small. It is expensive to add a story to an existing structure. In all probability Blackacre Estate would never have been expanded upward. What is real, however, is the collector owner's sense of security.

The mini-TDR concept is not relevant to vegetation since present laws do not limit the height of private vegetation. There is nothing to "trade" for new restrictions. Therefore, vegetation controls would have to be separately designed and implemented.

Common law backup and mini-TDRs are not included in the commentary to the legislation that follows as they would complicate it unnecessarily. Readers are urged to study the comments since they contain material not found elsewhere.

An Ordinance Concerning Solar Energy, Providing for Recordation of Solar Collectors, and Affording Solar Access Protection for Recorded Collectors

Introductory comment. This prototype ordinance protects access to sunshine by having the solar collector owner apply to record the existence of his or her collector in the public land records. If a recordation is granted, the law then protects the collector from unreasonable shading. Although recordation is considered on a case-by-case basis, decisions are made within the context of a citywide solar access plan, which helps to mitigate the "first-person-to-develop-wins" nature of this legislation (see Chapter 9). As explained in the introduction to Part III, ELI generally dislikes approaches based on priority of development. Their use should be limited to areas where few persons are expected to use solar energy systems. Although there are elements of zoning in this strategy, it does not rely on zoning to the same extent as does the prototype legislation in Chapter 5. A modified form of this legislation could also be adopted by counties, or by states to require or authorize local governments to protect solar access.

The prototype ordinance recognizes that in cities some areas are more suitable for solar collectors than others. Because protection of solar access restricts nearby development, city planning officials would be directed to divide the municipality into two categories of "overlay" zones. Overlay Zone One would include those well-settled residential and commercial areas within which nearly all lots have structures that are already close to the heights or densities presently allowed and where little further development is expected or likely. Overlay Zone Two would include areas that are not generally built up to their full permitted height or density. In such areas the protection of solar access could put a heavy economic burden on properties near solar buildings. A city's comprehensive plan would be important in determining the boundaries of these overlay zones, and the plan should be amended to incorporate solar access protection as a goal.

In both overlay zones, collector owners would apply for solar access recordations by the same process. They would provide information on the exact location of the proposed collector and the percentage of energy that it is expected to contribute to a particular use. The legislation spells out detailed steps for notifying neighbors; if the neighbors register objections to the recordation, a public hearing would be held. (The impracticality of having a large number of hearings reduces the usefulness of a case-by-case approach such as this to areas where only a few solar installations are expected; where large numbers of installations can be expected, legislation using a planned approach, such as solar envelopes, is preferable.) Solar system owners are not required to record their collectors: they may decide to accept the risk of shading or to negotiate private agreements with their neighbors. A recordation may be applied for either before or after a collector is installed.

The different natures of the two overlay zones would be considered when the relevant government agency decides whether to grant a recordation. Recordations would be somewhat easier to obtain in an area designated Overlay Zone One than they would be in Zone Two. In a Zone One, someone (including the city) wanting to prevent a recordation would have to prove that the property should have been classified Zone Two (Zone Two standards would then apply), that he or she has present plans for development that would shade the collector, or that the collector would unreasonably restrict the future development of the objector's land.

In a Zone Two, slightly different factors determine whether a recordation would be granted. Basically, a recordation could be obtained only if the collector owner could show that the collector would not be shaded if nearby properties were developed as much as was currently permitted or likely to occur. Unlike the situation in Zone One, it is the collector owner who would bear the burden of proof.

Once a solar collector recordation was granted, the city could not issue any development permits that would result in the shading of the recorded collector, and neighbors would have to keep their vegetation from growing so that it shaded a collector. When a neighbor could prove that vegetation existed before a collector was recorded, the collector owner would have to reimburse the neighbor for the cost of keeping the vegetation trimmed.

A final section of the prototype legislation would allow the city to repurchase the rights granted by solar collector recordation if the city decides to permit development that would shade a recorded collector. This provision ensures that solar access protection will not freeze urban growth or redevelopment and that technological and societal changes can be accommodated. Rather than have the city rely solely on its eminent domain powers for the repurchase, the owners of recorded collectors would agree to be bound by arbitration in the event of repurchases.

Note: This prototype ordinance is not intended to be adopted in its present form. It is a longer law than will be required in many localities, and is deliberately comprehensive to serve as a reminder of the many possible issues that may be encountered. Many of the goals expressed in this ordinance could, instead, be included in administrative regulations. Much of the language could also be used in a state-level law or guidelines accompanying such a law. The ordinance format was selected because of its general appropriateness for solar access legislation. When brackets appear in the legislation they indicate either (1) alternative approaches or (2) places where local expertise must supply the appropriate name or number. The bracketed numbers are simply suggestions of what may be reasonable.

Enactment Clause with Purpose and Findings

See the alternative enactment clauses and accompanying commentary in the prototype legislation in Chapter 5.

SECTION 1 Short Title

This Ordinance may be cited as the "[city name] Solar Access Ordinance."

SECTION 2 Definitions

As used in this Ordinance, the following terms shall have the meanings shown:

(a) "Development permit." Any permit or authorization issued by the City as a prerequisite for undertaking any development. It includes permits and authorizations customarily known as building permits, zoning or rezoning permits, variances, special permits, plat approvals, or subdivision or large-scale development permits.

(b) "Solar access recordation." A recordation issued under the procedures set out in this Ordinance.

(c) "Direct solar energy." Direct radiant energy received from the sun.

(d) "Solar collector." A device, or combination of devices, structures, or part of a device or structure that transforms direct solar energy into thermal, mechanical, chemical, or electrical energy that substantially satisfies a structure's [or a swimming pool's] energy requirements [in accordance with adopted standards in regulations of the planning department or board of zoning appeals].

(e) "Collector surface." Any part of a solar collector that absorbs direct solar energy for use in the collector's energy transformation. It does not include such items as frames, supports, and mounting hardware.

(f) "Recorded collector." A solar collector for which a solar access recordation has been issued or for which a recordation application is pending.

(g) "Recorded owner." The holder of a solar collector recordation issued under this Ordinance.

(h) "Shaded." A solar energy collector is deemed shaded if vegetation or structures block the direct solar energy that would otherwise reach its collecting surface during the collector use period (as defined in Subsection (i)). Such insubstantial shadows as those cast by utility poles, wires, flagpoles, and slender antennas are not deemed to shade for the purposes of this Ordinance. The [board of zoning appeals] may adopt regulations, or rule on particular cases, to further define such insubstantial shading exempted from regulation.

(i) "Collector use period."

 (1) For solar space heating collectors, or for collectors used for both space heating and cooling, [9 A.M. to 3 P.M. compensated standard time or local solar time] during [the months of the local heating season].

 (2) For solar space cooling collectors, [8 A.M. to 4 P.M. compensated standard time or local solar time] during [the months of the local cooling season].

 (3) For solar hot water heating collectors, [9 A.M. to 3 P.M. compensated standard time or local solar time] year round.

 (4) For other solar applications [such as swimming pool heating], such periods as the [planning department or board of zoning appeals] shall determine in its regulations.

(5) In the case of any amendment to the collector use period, the collector use period as defined on the date of issue of the solar collector recordation shall apply. A recorded owner may apply for a new recordation granting the new period of protection if a longer collector use period is promulgated.

(j) "Registered lessee." Holder of a lease registered in the [local land records].

Comment on definitions. "Solar collector" as defined here does not include direct applications of daylight, for there is no energy conversion in such applications. While the ordinary use of windows for daylight illumination is not worthy of the kind of strict solar access protection given under this ordinance, some forms of light collectors (those that require direct sunlight, supply most of the lighting needs of a building, and meet the "contributes significantly to a structure's energy supply" criterion) may deserve protection. If this is desired, the collector definition should be changed to reflect this and regulations should clarify what energy contribution requirements have to be met. The term "contributes significantly" is deliberately vague. ELI believes that any serious use of solar collectors merits some access protection. But neighbors of solar collectors should not have their property rights reduced to protect access for a solar toy or a small south-facing window. Localities may wish to set a minimum Btu contribution in their definition of collector, but such an approach could put a heavy burden of proof on collector owners. Although the definition of collector is broad, this ordinance does not protect trivial uses of solar energy—Section 6(a)(3) requires the owner of the collector to state the percentage of energy that the collector is expected to provide for any particular use. Recordation will be denied if the amount is lower than that required in regulations [Section 6(b))].

"Shaded" includes all shading during the collector use period except certain insubstantial shadows exempted by enumeration in the ordinance or by regulation. An alternative but unsatisfactory approach is to allow for loss of a certain percentage of the otherwise available solar energy. Massachusetts bills make shading an "impermissible interference" if the blockage reduces the total solar energy otherwise available to a collector by more than 10 percent (S. 304, 1977; H.B. 3842, 1978). Such an approach unfortunately creates unsolvable problems of proof. How does one know how much blockage amounts to 10 percent of otherwise available solar energy? It would be necessary to account for such factors as actual weather conditions (e.g., early morning fog), structures, and vegetation existing before the collector went in, and actual topography. A computer would be needed to figure whether particular shadows are permissible. Proving shadows to be impermissible would place an unreasonably heavy burden of proof on the collector owner.

"Collector use period" as specified here could be adjusted according to local conditions, such as latitude, weather, and local heating/cooling periods. The hours of protection could be longer or shorter and should be translated into compensated standard time using the method in Appendix A. Protection of access during the middle two-thirds of the day will provide roughly 80 percent of the solar energy available, which is a reasonable objective of protection.

"Registered lessee" is used here to permit actions against the person who has record control of activities on land, such as the holder of a long-term lease registered on the local land records.

SECTION 3 Applicability of Ordinance

(a) Governmental organizations not under the jurisdiction of the City (such as the State and Federal governments) may elect to enjoy the benefits of solar access permits under this Ordinance if they also consent to be bound by its restrictions. An organization not having so consented may neither apply for a solar access permit nor protest the application of another.

(b) City property shall be subject to this Ordinance. The City may apply for recordation under Section 6 and may protest any application for a solar access recordation under Sections 6 and 7. In such actions the City shall be represented by [appropriate agency].

(c) When the provisions of this Ordinance conflict with any other laws or regulations relating to land use, the requirements that most restrict development shall apply.

Comment on applicability. It is useful for the state and federal governments to consent to be bound by solar access legislation. This provision provides an incentive for these governments to follow local land use rules protecting solar access. The ordinance makes it clear that its benefits go only to those who will be bound by it.

Subsection (b) allows the city itself to apply for recordation or on its own motion to act to protest solar access recordations. It may wish to do so to protect neighbors who are unable or unwilling to protect their own interests, or to ensure that permits will not be granted contrary to its overall land use policies.

SECTION 4 Comprehensive Plan

The [name of appropriate planning body] is directed to:

(a) Assess the likely rate, pattern, and extent of solar energy usage within the community as a whole and within its districts over a period of [25] years.

(b) Prepare a map showing the Solar Access Zones established in accordance with Section 5 of this Ordinance. A copy of the map, together with any amendments adopted from time to time by [the city council], shall be provided to the [building office].

(c) Prepare an amendment to the City's Comprehensive Plan according to the procedures contained in [ordinance or statute setting out procedures to be followed in adopting the comprehensive plan]. The amendment shall encourage the use of solar energy in the City, shall incorporate the assessment and map developed under Subsections (a) and (b), and shall, to the extent feasible, be in harmony with other goals expressed in the Comprehensive Plan.

The [planning body] is directed to complete the revision of the Comprehensive Plan and the division of the City into Solar Access Zones within [six months] following the adoption of this Ordinance. The completed solar access map and revision of the Plan shall be submitted to the [city council] for review and adoption. Pending completion of this review and adoption, applications for solar collector recordations may be made but shall not be processed.

Comment on comprehensive plan. This prototype ordinance adopts what is at once the conservative position and the modern position: that a comprehensive plan that is independent of the zoning ordinance is needed to comply fully with zoning enabling legislation (see Chapter 3). By lending weight and logic to the divisions called for in Section 5 of this ordinance, a plan will help a community that uses solar access controls to protect itself against legal challenges, particularly those questioning whether the ordinance only "incidentally" benefits solar system owners (see Chapter 3). If a locality has no comprehensive plan but wishes to adopt solar access protection similar to that described in this prototype legislation, it should still perform steps (a) and (b), but instead of (c) it should prepare a new comprehensive solar plan based on (a) and (b).

SECTION 5 Solar Access Zones

The Solar Access Zones shall be mapped as follows:

(a) **Overlay Zone One.** An existing zone shall be designated as in Overlay Zone One if at least [80] percent of the lots located within it have structures now developed to [within one-and-one-half stories OR: within [15] feet] of the current maximum permitted height.

(b) **Overlay Zone Two.** An existing zone shall be designated as in Overlay Zone Two if it fails to meet the criteria for inclusion in a Zone One area or, notwithstanding the preceding subsection, if in the opinion of the [planning body] a Zone One designation would cause great economic hardship or be clearly inconsistent with the planned or reasonably anticipated development of any particular area, even though this is not reflected in current zoning.

When appropriate, existing zones may be divided so that they are partly in Overlay Zone One and partly in Overlay Zone Two.

Comment on solar access zones. This section directs the appropriate planning body (usually the planning department or commission) to divide the city into two types of overlay zones. Areas designated Zone One (not necessarily contiguous) are where nearly all lots have structures built up to, or close to, the legally permitted height, so solar collectors will impose less of a burden on neighbors. In developing or changing areas (Zone Twos), however, the burden might be substantial, and collector recordations should be correspondingly harder to obtain. Limiting the easy granting of recordations to areas where the burden will be relatively small, while making them hard to obtain elsewhere, mitigates constitutional problems discussed in Chapter 3.

Defining the zones involves judgments on the part of the city staff. It is not possible to write an ordinance anticipating every relevant detail. Allowance is specifically made, however, for planned anticipated development not yet reflected in current zoning. The planning staff should consider current zoning in determining the shapes and sizes of the solar access zones, but zoning should be only one factor, not the controlling one. For example, low-rise areas near the line between two zones may not be well suited for solar development if high-rise buildings are permitted to the south. Similarly, areas that slope steeply to the north are unlikely candidates for Zone One. Rezoning underlying zones to lower densities should be considered if planners discover that nearly all structures within an existing zone are about the same height but significantly shorter than the maximum allowed height. A Zone One designation would then be possible.

A simpler alternative to dividing a city into solar access overlay zones is the method proposed in the state of Massachusetts (S.B. 304, 1977), which divides cities on the basis of whether the area is zoned for more or fewer than four stories (Sections 23 and 24). This avoids the need to determine the nature of existing and planned development. The appeal of this simplicity may, for some communities, outweigh the advantages of the approach proposed here. But using present zoning as the only relevant factor has problems: it ignores the actual and anticipated development of an area and may thus restrict solar use more or less than is necessary. If such a method were to be applied, a locality might choose a different dividing line (such as two stories or six, or—even better—a height standard), as appropriate to local conditions.

SECTION 6 Application for Recordation of Solar Collector

(a) Any property owner or lessee, or agent of either, may apply for a solar collector recordation from the [board of zoning appeals or other planning agency]. The application shall be in such form as the [recordation agency] may prescribe, but shall, at a minimum, include the following:

(1) The applicant's name and address, and the owner's name, address, and the [local customary legal designation] of the property where the proposed collector is to be located.

(2) A statement by the applicant that his collector is already installed or that it will be installed on the property within [one year] following the granting of the recordation.

(3) An estimate of the quantity and percentage of energy that the collector is expected to provide for its particular use (space heating or cooling, water heating, etc.) and an explanation of how these figures were computed.

(4) A statement that the collector site will be shaded during

the year only by vegetation and structures specifically listed on the application.

(5) The proposed size and location of the collector, its orientation with respect to south, and its slope from the horizontal, shown clearly in a drawing or in words.

(6) An explanation of how the applicant has done everything reasonable, taking cost into account, in designing and locating the collector in a manner to minimize the impact the collector will have on the development of nearby properties.

(7) A separately signed and dated statement that the applicant agrees to submit to binding arbitration concerning the value of the recordation and the damages to be paid if the City diminishes the rights protected by the recordation following procedures in Section 10 of this Ordinance.

(b) If the application on its face complies with this Ordinance, and if the proposed collector is to provide as much energy for particular uses as the [municipal legislature or recordation or other agency] shall require in standards or regulations that it shall promulgate, then the [recordation agency] shall accept the solar collector recordation application and notify the applicant. The applicant shall, in the manner set out below, notify all property owners on whose property a structure legally permitted by existing height regulations could be built that would cast a shadow on the applicant's collector during the collector use period, and all property owners on whose property a [70]-foot tree could be located that would cast a shadow on the applicant's collector during the collector use period. The applicant must then tender proof to the [recordation agency] that the following actions have been taken:

(1) That a certified letter, return receipt requested, has been mailed to each property owner and registered lessee shown on the list provided to the applicant by the recording agency. Alternatively, the letter may be personally delivered if a signed receipt is obtained. The letter shall be on a form supplied by the [recordation agency] and shall contain, at a minimum, the following information: the name and address of the applicant; that an application for a solar collector recordation has been filed; that the recordation, if granted, may affect the rights of other

property owners to develop their property and may in addition impose on them duties to trim some vegetation at their own expense; the advisability of obtaining photographic proof of the existence of trees and large shrubs; the times and places where the application may be viewed; telephone number and address of a city department that will provide further information; and that any adversely affected person may object to the issuance of the recordation by a stated time and date, and how and where the objection must be made.

(2) That a sign has been posted for [30] days on the property where the collector is proposed, with the [recordation-issuing agency's] telephone number on it and the words "SOLAR COLLECTOR RECORDATION PENDING." The sign should be clearly readable from the road or street.

(c) If no objections are filed within [30] days following the date the final certified letter is mailed or the property is posted, whichever is later, the [recordation-issuing agency] shall issue the solar collector recordation.

(d) If any adversely affected person or the City files a written objection with the [board of zoning appeals or other appropriate body] within the specified time, and if the objections still exist after informal discussions among the objector, appropriate city staff, and the applicant, then a hearing date shall be set and a hearing held in accordance with the provisions of Section 7.

Comment on application for a recordation. The ordinance provides for elaborate procedures to protect the rights of nearby property owners and registered lessees whose property may be burdened. Actual notice is given to those owners and lessees who presently may place a structure on their property—whether they have actually built such a structure or not—that could shade the applicant's collector. These provisions acknowledge that the development rights of the nearby property owners and lessees could be diminished by the granting of the solar access permit, and that they must be given an opportunity to object and to be heard. Federal and state constitutions may not require procedural safeguards that are this stringent. They are recommended, however, because of the novelty of solar access laws and because political opposition to them may be less if voters realize that there is a mechanism for protecting their interests.

The recordation agency must promulgate regulations specifying the amount of energy that a collector must supply for a particular use if it is to merit access protection. It is recommended that the collector replace at least 20 percent of the energy required for the particular use (space heating, space cooling, or water heating—perhaps more for water heating) on a main structure during the collector use period. The recordation agency should also promulgate regulations for minimum energy contributions for other uses of solar energy, such as process heat and photovoltaics. In addition, it could also promulgate rules that specify what, if any, recordations may be obtained for the benefit of accessory structures.

The collector's minimum contribution to a particular energy use could be set in terms of Btus or Kwhs instead of a percentage basis. Use of a percentage criterion encourages energy conservation measures like insulation and storm windows more than a Btu or Kwh criterion. If south-facing windows are to be included in the structure, a somewhat lower minimum contribution figure (10 to 15 percent) might be desirable. The recordation agency should be given the power to promulgate regulations for applications other than solar heating and cooling.

As contractors who build improvements often assume responsibility for obtaining all necessary permits, a consumer might incorrectly assume that a solar collector recordation has been obtained. A community may wish to require contractors to obtain the recordation as a prior condition to installation since contractors who installed many collectors would be familiar with the law and better able to comply than buyers. If building permits are granted prior to solar collector recordations, they should prominently state that they create no solar access protection.

SECTION 7 Hearing Procedure

(a) The [board of zoning appeals or whatever agency is specified in Section 6(d)] shall send notice of the hearing on the recordation application to the applicant and to all persons who objected to issuance of the recordation. The hearing shall be held according to the rules of the [board], shall be on the record, and open to the public.

(b) If the application is for a collector to be located in Overlay Zone One, the objecting party shall be given the opportunity to demonstrate, under the criteria in Section 5, that the area has clearly been improperly classified Zone One. If the objecting party is able to show such an improper classification, the hearing shall proceed under the standards in Section 7(e). If the objector is unable to show such improper classification, the hearing shall proceed under the standards in Section 7(c).

(c) In a hearing on an application for an area properly classified as in Overlay Zone One, the [board] shall consider the matters required for applications set forth in Section 6(a) on which the applicant shall bear the burden of proof and the following factors on which the objector shall bear the burden of proof:

(1) A showing that the objector has present plans to build a structure that will shade the proposed collector site. Present plans may be shown by an application for a building permit, an expenditure of at least [$500] on planning or designing such a structure, or any other clear and convincing proof that substantial progress has been made toward planning or constructing a structure that would shade the proposed collector.

(2) A showing that the proposed collector would unreasonably restrict the development of presently underdeveloped property. Property is considered to be unreasonably restricted if a structure cannot be erected on the buildable area to within [one-and-one half stories] of the currently maximum permitted height [OR, depending on language used in Section 5(a), to within 15 feet of the current maximum permitted height].

If the objector is unable to prove either of these circumstances and the applicant makes the showings required by Section 6(a), the [board] shall approve the recordation. If the applicant has failed to list all structures or vegetation shading the proposed collector location in his application, the [board] may approve the recordation while adding the omitted shading structures or vegetation as exemptions from this Ordinance. If the objector shows that an unconditional approval of the application would prohibit his or her reasonable present plans for development, the [board] may approve the recordation adding the objector's plans for development as exemptions; if the objector then fails to develop within a reasonable time, the applicant may apply to the [board] for a removal, in whole or in part, of these exemptions. If the objector shows that an unconditional approval of the application would unreasonably restrict development of the objector's presently underdeveloped property, the [board] may approve the recordation adding such exemptions as are necessary to allow for reasonable development of the objector's property.

(d) If the application is for a collector to be located in Overlay Zone Two, the applicant shall be given the opportunity to demonstrate that, under the criteria in Section 5, the area has clearly been improperly classified as Zone Two. If the applicant is able to show such an improper classification, then the hearing shall proceed under the standards in Section 7(c). If the applicant is unable to show such improper classification, the hearing shall proceed under the standards in Section 7(e).

(e) In a hearing on an application for an area properly classified in Overlay Zone Two, the [board] shall consider the matters required for applications set forth in Section 6(a) and the following factors, on which the applicant shall bear the burden of proof:

(1) Whether the area that would be burdened by the proposed collector recordation is unsuitable for development because it is an historic district, flood plain, parkland, highway or other publicly owned property that is sufficiently wide to make shading of the collector impossible by a structure under current zoning.

(2) Whether the collector would be unshaded even if other properties are developed to the full extent currently permitted and likely to occur.

The [board] shall approve the recordation if the application can show either that the area burdened is not suitable for development under paragraph (1) or the collector would not be shaded under paragraph (2), and can make the showings required by Section 6(a). Notwithstanding that there will be some shading if full development is allowed, the [board] may approve the recordation, provided shading by development up to the currently permitted extent is exempted from this Ordinance.

(f) A losing party may appeal the determination of the [board] to [name of court of general jurisdiction in the city].

Comment on hearing procedure. In Overlay Zone One, a ground for objecting to the issuance of a solar collector recordation is that the objector could not develop property up to the heights enjoyed by neighbors. A city could let this objection block the issuance of a recordation only in instances where the city is unwilling to grant a variance allowing more density on portions of the objector's lot where construction would not result in shading of the collector.

In Overlay Zone Two, the ordinance was originally written to make it more difficult to obtain a permit. Because of the repurchase provisions in Section 10, however, it was decided to record collectors that are unlikely to impose hardships on their neighbors, while allowing the repurchase of the protected access if it later turns out that development becomes desirable. It seems fair to allow recordations to be issued where it is unlikely that structures will be developed to their maximum permitted size, since areas are sometimes zoned for much more intensive development than is really expected.

SECTION 8 Entry of Recordation into Register

When a solar collector recordation is granted, the [recordation agency] shall notify the [land title recording office] of the owner's name, address, and [the local customary legal designation] of the property where the recorded collector is to be located together with any special exceptions or exemptions from the usual effects of a recordation. The Recorder shall enter this information in a register and plot the recorded collector location on a map kept for this purpose.

Comment on entering recordation in register. An alternative to Section 8 is to note the permit on the title of all properties owned by persons given actual notice by letter and limit the effect of the permit to these properties. (See Mass. S. 304, 1977, § 24(b).) Although it would involve considerable paperwork, putting restrictions on titles to property gives the clearest notice of restrictions to subsequent purchasers. A look around the neighborhood could often show whether there are solar collectors that may require sunlight crossing above property someone wants to buy, but most people would not think to look for them. And passive-type collectors may incorrectly be assumed to be ordinary windows, or roof-top collectors may be screened from view. Some notice in the public land records is needed; the notice provided here should be adequate, but some localities may prefer to give more.

SECTION 9 Effect and Enforcement of a Solar Collector Recordation

(a) No City department shall issue any development permit purporting to allow the erection of any structure that would shade a recorded collector or a proposed collector site if a recordation application is pending or granted but the collector is not yet installed, except to the extent that the exemptions in the solar collector recordation itself would allow. [Development-permit issuing department] shall consult the map showing issued solar collector recordations and shall consult the [solar collector recordation-issuing agency] concerning any pending applications before issuing development permits. Whenever it appears that granting a development permit might result in a structure that would shade a recorded collector, the department concerned shall give notice by certified mail, return receipt requested, to the recorded collector owner. Any development permit that would result in the shading of a recorded collector is void except to the extent that it authorizes work that would not shade the collector. The recorded owner may appeal the grant of a development permit that purports to authorize a shading structure either by such administrative means as are available or to [appropriate court of general jurisdiction].

(b) Pending the resolution of his appeal, the recorded owner shall be entitled to a preliminary injunction to stop construction of such part of a structure as may shade the recorded collector. The party proposing to build the structure shall bear the burden of showing that the structure would not shade. If the court finds that the structure would shade, the injunction shall be made permanent and the development permit declared void to the extent the structure would shade. If the structure has already been built to the extent that it shades the recorded collector, the court shall order removal of the shading part if feasible or order the payment of damages using the measure of damages in Section 10. The recorded owner shall be entitled to the payment of appropriate damages, plus court costs, plus reasonable attorney's fees, as against either the City or the constructing party, or both.

(c) No one shall plant any vegetation that shades a recorded collector (or a recorded collector location if it is not yet installed) after receiving notice of the pendency of a collector recordation application or after issuance of a recordation. After receiving notice of a solar collector recordation or application, no one shall permit any vegetation on his or her property to grow in such a manner as to shade a recorded collector (or a recorded collector location if it is not yet installed), unless the vegetation is specifically exempted in the recordation.

(d) If vegetation is not trimmed as required or is planted contrary to Section 9(c), the recorded owner or the City on complaint by the recorded owner shall give notice of the shading by certified mail, return receipt requested, to the owner or registered lessee of the property where the shading vegetation is located. If the property owner or lessee fails to remove the shading vegetation within 30 days after receiving this notice, an injunction shall be issued, upon complaint of the recorded owner or the City to [court of general jurisdiction]. The injunction shall order the property owner or lessee to trim the vegetation, and the court shall order the violating property owner or lessee to pay any damages to the complainant, to pay court costs, and to pay the complainant reasonable attorney's fees.

(e) If personal jurisdiction cannot be obtained over either the offending property owner or registered lessee, the City may have a notice listing the property by owner, address, and [legal description], published once a week for four consecutive weeks in a newspaper of general circulation within the City, giving notice that vegetation located on the property is in violation of this Ordinance and is subject to mandatory trimming. The City shall then have the power, pursuant to court order, to enter the property, trim or cause to have trimmed the shading parts of the vegetation, and add the costs of the trimming, court costs, and other related costs to the tax liability attached to that property [OR: as a lien against that property].

(f) If the City is the property owner or registered lessee and has failed to trim vegetation shading a recorded collector, the recorded owner may notify the City [specific department] by certified mail, return receipt requested, asking the City to trim the vegetation. If the City fails to do so within 30 days after receiving this notice, the complaining recorded owner may bring suit for an injunction requiring trimming and may recover damages, court costs, and reasonable attorneys' fees from the City, if the City is found to be in violation.

(g) Where the property owner or registered lessee contends that particular vegetation is exempt from trimming requirements, the burden of proof shall be on the property owner or lessee to show that an exemption applies to the particular vegetation.

(h) Notwithstanding any other provision of this Section, the owner of a recorded collector may allow any person to build a structure otherwise permitted by law or let any vegetation grow in a manner to shade a recorded collector. A signed, notarized document setting out the extent of the permitted shading may be accepted as proof of consent by a [city development permit-issuing office] in issuing a development permit that would otherwise be unlawful under this section.

(i) A property owner or registered lessee shall be entitled to reimbursement from the registered owner for all reasonable costs of keeping vegetation trimmed as required under this Ordinance, provided the property owner or lessee can show that the vegetation in question was in place on his property before he received notice of the solar collector recordation application.

(j) A solar collector recordation becomes void and shall be removed from the register and map, if a functioning collector is not installed within [one year] after issue of the recordation.

(k) A solar collector recordation also becomes void and shall be removed from the register and map, if the collector is removed and not replaced, or not used, for [two years], excluding time spent on repairs or improvement.

(*l*) The transfer of title to property shall not change the rights and duties provided for in this ordinance.

Comment on effect and enforcement of solar collector recordation. A solar collector recordation has some of the characteristics of a private easement, although for structures it is largely administered by the local government. The burdened property owners have a duty to trim shading vegetation. Since the police power has not been traditionally used to regulate privately owned vegetation in this manner, amendments to state enabling legislation may be necessary. The ordinance might have been written so that collector owners could enter their neighbor's land to trim the shading vegetation themselves. Easements sometimes allow similar privileges. In this case it would be unwise as it would cause arguments over whether trimming was done properly or was required at all.

Another possibility is to declare shading vegetation a public or private nuisance, since a city could also exercise its *in rem* jurisdiction (when *in personam* jurisdiction was not possible) to attach the property with shading vegetation, sell the property, and pay monetary damages from the proceeds. This seems more drastic than is necessary.

Section 9(g) places the burden on the owner of the vegetation to prove that it was there before the collector recordation. A simple way to do this would be with a notarized, dated photograph of the vegetation. When notice is given that a solar collector recordation application is pending, it could be required that the notice include language advising people to prepare such proofs. Such proof would also be useful under Section 9(i), which requires the collector owner to compensate neighbors for the cost of trimming in situations where the tree or bush predated the notice of the application for solar collector recordation.

The ordinance as written does not exempt preexisting vegetation, but only preexisting shadows on collector sites. This does have the disadvantage that it imposes a duty to trim on the vegetation owners. Some cities may prefer to exempt preexisting vegetation instead. The burden of proof should remain, however, on the owners of the vegetation. It would be too difficult for the collector owner to show that vegetation was not there.

SECTION 10 Repurchase of Recordation

(a) Any recordation issued pursuant to this Ordinance is subject to the repurchase requirements of this Section. In areas included in an Overlay Zone One, however, a recorded collector owner may require the City to prove that the neighborhood has changed in character before a repurchase may be made. A change in character means that a neighborhood no longer fits the description in Section 5(a) and now is best described as a Section 5(b) situation.

(b) Solar collector recordations shall contain a condition clearly stating that at any time the City may repurchase all or part of the rights granted by the recordation, with the qualification listed in Subsection (a). At the time of the recordation application, the applicant shall agree in writing to follow a procedure for valuing the recordation at the time of any buy-back as follows:

(1) **Arbitration.** In Section 6(a)(7) of this Ordinance the recorded owner agrees as a condition of obtaining a recordation to submit to binding arbitration concerning the value of the recordation and the damages to be paid if the City diminishes in whole or in part the rights protected by the recordation. The arbitration shall be held pursuant to the rules of the American Arbitration Association, as amended from time to time. The presumptive measure of damages shall be the greater of the present net worth of the collector and associated equipment or the discounted present worth of the difference in energy costs with and without the collector for the remaining life of the collector and associated equipment, plus the cost of removing the collector (and repairing or putting on a new roof, siding, etc. if needed), plus the cost of providing mechanical equipment properly sized to perform the task formerly assumed by the collector and associated equipment, plus the costs to the recorded owner of the repurchase proceedings. The burden of showing the amount of these presumptive damages shall be on the recorded owner. The City shall be given the opportunity of showing by clear and convincing evidence that the actual damages are less than would be calculated by this formula and the extent of the actual damages.

(2) **Eminent Domain.** If for any reason binding arbitration is not legally possible, the City must use its normal eminent domain procedures.

(c) In either Subsection (b)(1) or (2) situations, if a private individual stands to benefit from the repurchase of a solar collector recordation, the City may require that individual to reimburse the City for the cost of the repurchase as a prior condition to granting a development permit that would authorize what would have been an unlawful shading structure.

Comment on repurchase of permits. The purpose of Section 10 is mainly to allow the normal redevelopment and rezoning of a community. The arbitration provision is included as the transaction costs of this approach would be less than for eminent domain. Mediation, which is less formal than even arbitration, is another possibility. Eminent domain (see Chapter 11) is a cumbersome process that should be avoided. Although the police power may be broad enough to allow solar access protection to be blocked later without compensation to collector owners, such an approach seems unfair and could discourage investment in solar equipment. Particularly in neighborhoods where new development is not expected, people should be entitled to rely on continued access to sunshine. Therefore, the proposed legislation would make it difficult for the city to repurchase solar access rights in such areas. There are reasons for not making repurchase too easy: people do not install solar collectors for purely financial reasons, and the aesthetics of a structure may be impaired if a solar system is made dysfunctional. Adequate compensation is therefore difficult to provide.

[Further SECTIONS may provide severability, effective date, safety, and amendment clauses as desired.]

Comment on further sections. In addition to the matters listed, if the local planning department or board of zoning appeals does not have rulemaking power, it would be wise to allow for a power to provide regulations to elaborate on the provisions of the ordinance and make their applications more intelligible to the average citizen.

8
Nuisance and Trespass Law

Common Law Approaches in General

Because no ordinances or statutes existed that dealt specifically with solar access, early researchers of the subject turned to the common law. They probed and prodded this amorphous body of law, often applying considerable intelligence and imagination in their efforts. Solar collectors did not exist when the great body of common law was developed; therefore, researchers' efforts focused on fashioning appropriate analogies.

The following common law theories have been suggested as mechanisms for protecting solar access:

- □ Public and private nuisance law
- □ Trespass law
- □ Prescriptive easements for light (including the ancient lights doctrine)

These theories have been put forward with either of two hopes in mind: (1) that courts will extend their application to cover disputes over solar access; or (2) that legislatures or local governments will use common law theory as a model for new laws. This chapter examines nuisance law and the closely related common law of trespass. Prescriptive easements are covered in Chapter 9. The advantage of applying old law to new situations is that courts and citizens feel comfortable with laws they are used to. It also makes sense to avoid writing new laws if they are unnecessary.

ELI does not, however, recommend the extension of common law approaches to solar access protection. To work effectively, the common law would need to be so greatly modified that it would be less confusing to begin with a new law. These common law approaches deserve attention, however, not only for their possible merit, but also because of their potential to create problems in solar access cases. Where legislatures fail to enact specific access laws, it is conceivable that courts could apply some common law theories.

Nuisance Law

It is generally understood that a public nuisance is "virtually any form of annoyance or inconvenience interfering with common public rights" and that a private nuisance is an interference with an individual's right to use or enjoy his land.[1] As opposed to a private nuisance, a public one must infringe on the rights of some appreciable portion of the community, not just on the rights of one person.[2] Both public and private nuisance laws have been proposed to protect solar access.[3]

At first glance, these definitions seem broad enough to encompass solar access disputes. For example, the owner of a $10,000 solar space heating system may call a neighbor's new garage a "nuisance" if it shades his collectors. But the particulars of nuisance law raise serious questions about its applicability. Nuisance law is extremely complex. One expert has complained of both public and private nuisance law:

1. William L. Prosser, Law of Torts, 4th ed. (St. Paul: West Publishing Co., 1971), § 87, p. 572.

2. William Q. de Funiak, Handbook of Modern Equity, 2d ed. (Boston: Little, Brown and Co., 1956), § 32, pp. 59–60.

3. The Minnesota Energy Agency has proposed legislation that would amend Minn. Stat. § 116H.127 (1976) to declare *private* nuisances to exist when solar energy systems are shaded by trees or shrubs between 9:00 A.M. and 3:00 P.M. The solar system would have to be capable of generating over five million Btu's a year to qualify for protection, and its performance would have to be significantly impaired by the shadows. It was proposed that the offense be a private rather than a public nuisance, because only those benefiting from the solar system are actually harmed, and because the Minnesota Statutes do not discuss public nuisances. Minnesota Energy Agency, *Legislative Options for Encouraging Solar Energy Use in Minnesota: Executive Summary* (13 December 1977), pp. 12–13, 21 (available from Minnesota Energy Agency, 150 E. Kellogg Blvd., St. Paul, Minn. 55101). Public nuisance law is proposed more frequently. *See,* e.g., Cal. A.B. 2321, which was enacted 31 August 1978 and will be codified as § 25980 et seq. of Division 15 of the Public Resources Code; Del. S.B. 647 (1978). The town of Kiowa, Colorado, has an ordinance declaring shadows on collectors to be a public nuisance.

> There is perhaps no more impenetrable jungle in the entire law than that which surrounds the word "nuisance." It has meant all things to all men, and has been applied indiscriminately to everything from an alarming advertisement to a cockroach baked in a pie.[4]

Nonetheless, policymakers will want to have a cursory knowledge of those aspects of nuisance law that are relevant to designing solar access legislation.[5]

- ☐ If a community does not enact explicit solar access legislation courts might conceivably apply nuisance law to settle solar access disputes. Such applications could sometimes lead to irrational land use.
- ☐ Even if legislation (or an ordinance or a regulation) exists that deals with solar access, courts may fill in any gaps in the new law with nuisance law reasoning.
- ☐ Nuisance law approaches are being considered for legislation in several states and a nuisance-based law was enacted in California.[6]
- ☐ A nuisance law approach to the regulation of shadows from vegetation may be feasible in a low density residential setting.

Judicial Implementation

Public nuisance law. The word "public" would be the obstacle in attempts to get courts to declare shadows on collectors public nuisances. In nearly all jurisdictions, a nuisance is not public if it inconveniences only a handful of persons. For example, a polluted stream that harmed only a few riparian owners was found to be a private, not a public, nuisance.[7] A troublesome shadow clearly falls more heavily on the owner of the shaded collectors than on the general public. It is true that the public has a marked interest in the development of solar energy, but this is a rather indirect public impact compared to the impact required in traditional nuisance cases. A few

4. Prosser, Law of Torts, § 86, p. 571.

5. A distinct nuisance law issue will be claims that glare off solar collector surfaces constitutes a nuisance.

6. See note 3 above. In addition, both Colorado and New Mexico have seriously considered nuisance law.

7. Smith v. City of Sedalia, 152 Mo. 283, 53 S.W. 907 (1899). *Accord,* Hartung v. County of Milwaukee, 2 Wis. 2d 269, 86 N.W.2d 475 (1957), *rehearing denied,* 87 N.W.2d 799 (1958).

states, however, do recognize a public nuisance if there is an interference with a "considerable number of persons," regardless of whether a public "right" is involved.[8] Under this approach, a single tall building that shaded many solar installations might be termed a public nuisance even if the building predated the collectors. A public nuisance might also be found if the energy from a collector was shared by people owning separate lots. If those states with lenient definitions of nuisance want to ensure that nuisance law will—or will not—be applied to solar access cases, they should so legislate.

Private nuisance law. In jurisdictions that do not enact solar legislation, suits based on private grounds may be the only hope for the owner of a newly shaded collector.[9] Private nuisance law may seem a logical way to protect solar access since "the essence of a private nuisance is an interference with the use and enjoyment of land."[10] However, the mere interference with the use and enjoyment of land is an insufficient basis for a private nuisance action; there must also be a substantial interference with a legally protected right.[11] Spite fence cases are an exception in some states,[12] and at least one writer argues that they set a precedent that could be useful in right-to-light cases.[13] There is currently no right to the free flow of light in this country (see Chapter 9).

Other obstacles would also have to be overcome by the plaintiff in a private suit. If a nuisance action is to be successful, "the plaintiff cannot, by devoting his own land to an unusually sensitive use

8. City of Phoenix v. Johnson, 51 Ariz. 115, 75 P.2d 30 (1938); Pennsylvania Coal Co. v. Mahon, 260 U.S. 393 (1922); Rex v. Lloyd, 4 Esp. 200, 170 Eng. Rep. 691 (1802); Higgins v. Connecticut Light & Power Co., 129 Conn. 606, 30 A.2d 388 (1943); People v. Brooklyn & Queens Transit Corp., 258 App. Div. 753, 15 N.Y.S.2d 295 (1939), *aff'd,* 283 N.Y. 484, 28 N.E.2d 925 (1940).

9. *See* Franklin Gevurtz, *Obstruction of Sunlight as a Private Nuisance,* California Law Review 65 (1977): 94. The author argues that changing values suggest that courts should recognize a private-nuisance cause of action for obstruction of light and air. *See also* Dale D. Goble, *Solar Rights: Guaranteeing a Place in the Sun,* Oregon Law Review 57 (1977): 94, 125 n.127.

10. Prosser, Law of Torts, § 89, p. 591.

11. Ibid., § 87, p. 577.

12. Ibid., § 89, p. 598 n.97. A spite fence is one that has no utility and is erected out of malice.

13. Goble, *Solar Rights,* pp. 125–126. Courts have consistently refused to recognize a right to light. Recent cases include Cain v. American Nat'l Bank & Trust Co., 26 Ill. App. 3d 574, 325 N.E.2d 799 (1975), and Piccirilli v. Groccia, 327 A.2d 834 (R.I. 1974). Goble notes, however, that courts tend to beg the question when concluding there is no right to light. *Solar Rights,* p. 125.

. . . make a nuisance out of conduct of the adjoining defendant which would otherwise be harmless."[14] Is a solar collector an unusually sensitive use? It is difficult to predict a court's response. Because solar collectors are not yet widely used, it is certainly possible that their use might be deemed what courts call hypersensitive. On the other hand, Congress has declared a national policy favoring the solar heating and cooling of buildings,[15] and the use of solar energy devices is rapidly increasing.[16] For these reasons, courts may hesitate to declare solar collectors unusually sensitive uses. Another factor that may influence a court is the nature and use of the land around the site where the collector is installed. A collector on the side of a building in midtown Manhattan is more likely to be found a hypersensitive use than one installed on a home in a sparsely populated area of New Mexico. A legislature could, of course, specify that collectors are not a hypersensitive use.

Another drawback to the private nuisance approach may be the hesitancy of courts to find a use that is legislatively authorized by zoning laws to be a nuisance.[17] In conflicts over solar access, the building causing the offending shadow will almost always be in full compliance with zoning laws. But according to one expert on land use law, the fact that a use is consistent with local zoning laws in no way renders that use immune to a nuisance action. The right to bring a nuisance action exists apart from zoning, he says, and, technically speaking, neither is considered relevant in connection with the other.[18] This expert notes that while the zoning of an area is considered by the courts when they evaluate the balance of interests, it is but one among many factors considered and is not necessarily conclusive.

14. PROSSER, LAW OF TORTS, § 87, p. 579.

15. E.g., Solar Heating and Cooling Demonstration Act of 1974, § 2, 42 U.S.C. § 5501 (Supp. V 1975); Solar Energy Research, Development and Demonstration Act of 1974, § 2, 42 U.S.C. § 5551 (Supp. V 1975).

16. *See*, e.g., W. A. SHURCLIFF, SOLAR HEATED BUILDINGS: A BRIEF SURVEY, 13th ed. (Cambridge, Mass., 1977), p. 9.

17. Ralph E. Becker, Jr., *Common Law Sun Rights: An Obstacle to Solar Heating and Cooling?* JOURNAL OF CONTEMPORARY LAW 3, no. 1 (1976): 30.

18. NORMAN WILLIAMS, JR., AMERICAN LAND PLANNING LAW, 5 vols. (Chicago: Callaghan & Co., 1975), 5: § 153.11, p. 238. *See also* A. JAMES CASNER AND W. BARTON LEACH, CASES AND TEXT ON PROPERTY, 2d ed. (Boston: Little, Brown and Co., 1969), p. 1242.

19. WILLIAMS, AMERICAN LAND PLANNING LAW, 5: § 153.11, pp. 238–39.

Conclusions. Common law nuisance should not be relied upon because courts are unlikely to find either public or private nuisance law applicable to the shading of solar collectors. Since public nuisance law is extremely unlikely to be applied in the absence of legislation directing courts to do so, the following analysis applies only to the judicial application of private nuisance law.

A common law nuisance approach does have several attractive qualities:

- No legislation is required.
- The balancing of interests and choices of remedies (injunction, damages) provides flexibility. For example, a court will consider the ability of the defendant, while still making reasonable use of his land, "to reduce or eliminate the inconvenience to the plaintiff."[20] As public interests and needs evolve, so could a nuisance law approach evolve.
- Nuisance law has been applied to both structures and vegetation in conflicts not involving solar access.

Nevertheless, the disadvantages of a common law nuisance approach outweigh its advantages. The major drawback is that it would not integrate access protection with comprehensive planning but would inflict restrictions on random properties. Even the nature of the restrictions would vary randomly and bear no relation to rational land uses. The effectiveness of a nuisance approach would also be reduced by its complexity. Complexity can breed uncertainty, making potential collector owners reluctant to rely on nuisance law to protect access to sunshine.

Although due process requirements would probably be technically satisfied under nuisance law, the lack of clear advance understanding of rights and duties may generate conflicts. If courts took a balancing-of-interests approach, even greater uncertainty would prevail. Most people would probably be unwilling to invest in solar equipment without more assurance. There would always be a possibility that damages instead of an injunction against the shading would be awarded.[21] Damages would not satisfy persons who

20. Prosser, Law of Torts, § 89, p. 599.

21. Ralph Becker notes that about half the jurisdictions "balance conveniences." He cites Boomer v. Atlantic Cement Co., 26 N.Y.2d 219, 257 N.E.2d 870 (1970), where damages were given in lieu of an injunction to persons living in a suburb blighted by heavy pollution from the cement plant.

installed solar energy systems to gain control over their energy source, to conserve fossil fuels, or to avoid polluting the environment.

Conflicts could be expected if property owners were denied the right to develop their land in ways authorized or encouraged by land use laws. The resolution of disputes would be complicated and extremely burdensome on all parties. In a private nuisance suit, both parties would have to hire lawyers and perhaps endure long delays while waiting for court hearings and appeals. Delay often makes construction projects impractical if their financing is highly leveraged. If the collector owner seeks an injunction to stop construction while waiting for a court decision, he may be asked to post an expensive bond to cover any damages should he lose the suit.[22]

Legislative Implementation

Most states have used their police powers to enact one of two types of statutes dealing with nuisances. Some states simply adopt in statute form the common law of public nuisance, as established by judicial precedent. Others declare specific acts to constitute public or (sometimes) private nuisances. The uncertainty over whether nuisance law is applicable to solar access conflicts could be alleviated by a legislative declaration that, under certain conditions, shadows on solar collectors are public or private nusiances.[23] A legislative declaration that something constitutes a nuisance will generally be upheld by courts unless it can be shown that the declaration reflects an arbitrary or capricious judgment, and such a showing is difficult to make. There are some precedents for declaring vegetation public nuisances, such as ordinances saying that trees that obstruct motorists' vision at intersections are nuisances.[24]

22. Reviewer Donald G. Hagman noted at this point that we appear biased toward public controls and administrative tribunals. He correctly pointed out that several very respected legal commentators have greater faith in the promise of nuisance law in general. E.g., *see generally* Robert C. Ellickson, *Alternatives to Zoning: Covenants, Nuisance Rules, and Fines as Land Use Controls,* University of Chicago Law Review 40 (1974): 681–781. Dale P. Goble's law review comment favors nuisance law over zoning for solar access protection. *Solar Rights,* p. 128.

23. California has declared vegetation shadows on collectors to be public nuisances. Cal. A.B. 2321 (passed 31 August 1978) (reprinted in Appendix B). Colorado has rejected a measure based on public nuisance law. Colo. S.B. 38 (1976). Legislators were concerned about who would bear the costs of controlling vegetation and about unfair restrictions that might result from collectors situated near southern lot boundaries.

24. E.g., Stevens v. City of Salisbury, 240 Md. 556, 214 A.2d 775 (Ct. App. 1965).

Because of potentially harsh results, it is probably desirable to limit any legislative proposals based on nuisance law to shadows from vegetation. The reason for this is that owners of trees and bushes generally have less money invested in them than they do in structures, and that citizens are used to having structures regulated only by building codes and land use laws such as zoning. If legislation declaring shadows from structures to be nuisances is nevertheless enacted, it should be integrated with comprehensive plans so that important planning goals would not be thwarted. The amount of solar access protection granted should be adjusted to be compatible with such goals.

A new public-nuisance-type solar access law should give individuals the right to sue; otherwise, aggrieved persons might have to wait for a public official to bring an action if they could not prove that they suffered an injury different in kind and not just degree from that inflicted on the general public.[25] Proving this should be possible, but it would be simpler to eliminate this common law requirement.

Clearly drafted legislation that applies only to shadows cast by vegetation in low density residential areas, and that exempts specified energy-conserving vegetation such as windbreaks, could be reasonably effective. However, property owners still could not be sure how particular courts would interpret nuisance law. An ordinance or statute could try to anticipate each type of nuisance law technicality that a court might consider and specify how it applies to solar access. For example, the law could state that a collector owner would not have a right to have structures or trees removed if they shaded the collector at the time of its installation, in spite of the usual right to have a nuisance abated even if the nuisance existed before the competing use. Because of the complexities of nuisance law, the statute limiting its application would also have to be complex. And legislators may fail to anticipate all undesirable interpretations. Because a lawsuit could not be brought until an actual conflict existed, the predetermination of precise rights in individual cases would not be possible.[26] As case law on the subject evolved, there would be more

25. *See,* e.g., Reynolds Metals Co. v. Martin, 337 F.2d 780 (9th Cir. 1964); Boomer v. Atlantic Cement Co., 26 N.Y.2d 219, 257 N.E.2d 870 (1970).

26. Courts will seldom grant an injunction before a nuisance actually occurs. "What this means is not merely that the remedy in nuisance is unpredictable in advance: when it is invoked, the result is likely to be considerable social waste." WILLIAMS, AMERICAN LAND PLANNING LAW, 5: § 153.09, pp. 234–5. Furthermore,

certainty, but until this happened the use of solar energy systems might be discouraged.

Conclusions. A nuisance law approach to controlling vegetation shadows would be tolerable only if the remedies it provided were carefully shaped and limited so that they did not conflict with societal concerns reflected in land use laws, and only if the law applied to particular developmental conditions, such as suburban areas zoned for low densities or areas where new construction is not anticipated. It would also be necessary for legislation to specify where collectors could be placed and how large they could be (or which particular shadows, in terms of locations, would be nuisances).[27] Otherwise, properties would be unfairly and unequally burdened.

Not all problems raised by a nuisance law solution are amenable to resolution through careful drafting. For example, it would be stretching the concept of nuisance beyond its accepted meaning to try to use such a law to protect solar access that will be needed in the future. Though hybrid solutions could allocate responsibility in different ways, solar access controls based on nuisance law would tend to invite more judicial decisionmaking, while removing some control from legislative bodies and administrators. (A discussion of the advantages and disadvantages of judicial policymaking—or of the virtues of flexibility versus the dangers of uncertainty—are outside the scope of this study.)

On the whole, it is probable that using the word nuisance in solar access legislation invites more problems than it solves. If a community or state wants to protect solar access on a lot-by-lot basis, it may be better simply to prohibit certain shadows on collectors and provide appropriate remedies for when violations occur, than to label shadows nuisances. The same nuisance evokes a venerable body of complex law that offers no clear advantage but introduces many uncertainties.

courts have been extremely reluctant to allow administrators to decide that something is a nuisance and to terminate a use of private property without compensation. "Which, being translated, means that judges are men who can always tell whether a given activity is in fact a nuisance, whereas the personnel working in administrative agencies do not have the same powers of perception." Ibid., § 153.09, p. 236.

27. For instance, decisions should be made on whether south-facing windows should be granted legal protection from shading, whether only solar systems generating over a certain number of Btu's should receive legal protection, and whether a given sort of collector is reasonably sized and placed.

Trespass Law

Although the distinction between trespass and nuisance "has become wavering and uncertain,"[28] they are still two distinct bodies of law. A trespass action is based on an invasion of a landowner's exclusive possession of property.[29] Because an invasion is required, a trespass approach to solar access protection is illogical. Even the successful trespass suits that arose because of blasting involved the transmission of rocks or shock and sound waves over another's land.[30] Similarly, there have been decisions holding that invisible gases and microscopic particles may trespass.[31] However, courts have refused to find trespass when rays of light were cast on another's land.[32] It is even less likely that they would find the blockage of sunlight to constitute trespass. Since a shadow is merely the absence of light, it could hardly be said to invade. Nevertheless, it has been suggested that courts might apply trespass law to solar access disputes.[33]

Property owners do own the airspace directly above their land, but ownership does not splay out over adjacent parcels.[34] If a neighbor blocks the sunlight reaching your land by activities on his own land, he is not interfering with any of your possessory rights. Case law concerning trespass to vertical airspace (such as aviation cases) does not, therefore, apply.

For these reasons, existing trespass law offers no protection to solar collector owners. The question still remains whether it would

28. PROSSER, LAW OF TORTS, § 89, p. 594.

29. Ibid., pp. 594–95. A nuisance, on the other hand, is an interference with his use and enjoyment of real property. According to this distinction, nuisance would seem a more appropriate basis for an action for obstruction of light than trespass would. *See generally* Gevurtz, *Obstruction of Sunlight,* pp. 94–119.

30. *See* PROSSER, LAW OF TORTS, § 78, pp. 513-514.

31. Gregg v. Delhi-Taylor Oil Corp., 162 Tex. 26, 344 S.W.2d 411 (1961) (gases); Martin v. Reynolds Metals Co., 221 Or. 86, 342 P.2d 790 *cert. denied,* 362 U.S. 918 (1959) (gases and microscopic particles).

32. Amphitheaters, Inc. v. Portland Meadows, 184 Or. 336, 198 P.2d 847 (1948). *But see* Martin v. Reynolds Metals Co., 221 Or. 86, 342 P.2d 847 (1959). Transmission of light has, however, been held to constitute a *nuisance. See* Shelburne, Inc. v. Crossan Corp., 95 N.J. Eq. 188, 122 A. 749 (1923); Hansen v. Independent School Dist. No. 1, 61 Idaho 109, 98 P.2d 959 (1939).

33. MILAN JOHNSON AND THOMAS F. TIEDEMANN, SOLAR ENERGY COMMERCIALIZATION AT THE STATE LEVEL: THE FLORIDA SOLAR ENERGY WATER HEATER PROGRAM, report to the Federal Energy Administration (Cape Canaveral: Florida Solar Energy Center, March 1977), p. 58.

34. PROSSER, LAW OF TORTS, § 13, pp. 69–70.

be advisable for legislatures to enact statutes declaring trespass law to apply to shadows on collectors. Although there are statutes defining liability for trespass, such as for trespassing animals, there has not been extensive legislative definition of what constitutes a trespass. If a legislative approach were attempted, it would obviously be impractical to prohibit all obstruction of light. It would thus be necessary to decide which shadows to forbid and/or which collector sites merit protection, and to specify times or angles of protection and criteria for permissible shading. Because nothing would be gained by calling shading a trespass and because it would be illogical to apply trespass law to shadows, trespass-based solutions to solar access do not merit legislative attention.

9
Approaches Based on Accidents of Time

When Americans dispair of finding an equitable and practicable way of allotting something desirable, we often fall back on the familiar "first-come first-served" principle. Thus, the first 50 persons to arrive at a federal campground get sites, and the first miner to stake a claim gets rights to the ore. And in some states with race-notice systems of land registration, the person who first records a purchase gets priority over earlier buyers of the same real property who failed to record and give notice.[1] It is not surprising that similar suggestions have been applied to the frustrating issue of solar access. Three in particular deserve attention:

- □ Prescriptive easements for light
- □ "Instant access rights" legislation
- □ Western water law analogy

All these approaches share an arbitrary quality: whether one has a right to light or a right to develop is an accident of time. They also share the central failing of all approaches offering protection on a lot-by-lot basis: unequal burdens are imposed on similar land parcels. In other words, although the restrictions on construction and vegetation on privately owned property might sometimes be no

1. *See,* e.g., OR. REV. STAT. § 92–640.

greater than those imposed on construction by height and setback requirements, in other circumstances these restrictions could be intolerable (depending on the location of nearby solar installations).

Prescriptive Easements for Light

An easement is the right to use part of someone else's property or to prevent the owner from making full use of it. Easements may be formalized either by private agreement (see Chapter 10) or by judicial recognition. This discussion deals only with the latter. The person holding an easement does not have any ownership rights in the property subject to the easement. An easement is "prescriptive" if it is acquired by long, uninterrupted, and uncontested use. In the case of easements for light, there is no right to go on the other person's property: it is therefore a "negative" easement because it only restricts the other landowner's enjoyment of his land. If there is no legislation that sets the conditions under which a prescriptive easement will arise, landowners must rely on the common law to create such easements. Courts in the United States, however, have uniformly rejected the idea that easements for light can be created through prescription.[2] This is true even if an economic loss (such as a decrease in property value) is suffered because of the reduced light.[3]

In Great Britain and parts of the Commonwealth, prescriptive easements for light are possible under the ancient lights doctrine. A modern legislative version of the doctrine gives property owners a right to continue to receive a certain amount of sunlight in their room if they have been receiving that light, unblocked, for at least

2. Parker & Edgarton v. Foote, 19 Wend. 309, 318 (N.Y. Sup. Ct. 1838). *See also* cases cited in notes 3 and 7 below, and the many examples cited in these cases. Easements implied by *necessity* are also possible, although they have not been used to provide access to sunlight. One writer notes that courts tend to mean absolute rather than reasonable necessity in such cases, but that new solar technologies may influence this point. Gail Feingold Takagi, *Designs on Sunshine: Solar Access in the United States and Japan,* Connecticut Law Review 10 (1978): 127–28. Proposed Oklahoma H.B. 1741 (1978) would explicitly prohibit prescriptive easements for access to sunlight, but would allow those created "by necessity." Unfortunately, the bill does not elaborate on when light easements would be a necessity. This approach is not recommended as a general solar access vehicle because of the uncertainty over when rights would exist, and because these easements implied by necessity would arise only when both parcels were once owned by the same person, and the circumstances at the time one parcel was sold showed an unarticulated intention to create an easement.

3. Cain v. American Nat'l Bank & Trust Co., 26 Ill. App. 3d, 325 N.E.2d 799 (1975).

twenty-seven years.[4] Because the ancient lights doctrine is the closest analogy in Anglo-American law to a right to sunlight for solar collectors, it has been extensively discussed in solar access literature and even recommended as an approach to protecting solar access.[5]

Solar energy supporters are probably fortunate that American courts have rejected the ancient lights doctrine: the doctrine requires actual receipt and use of light before a right to it occurs. One would have to risk purchasing a solar energy system long before gaining assurance of continued access to sunlight. Prescriptive easements would probably not, however, promote races to develop land first because of the long waiting period before any right accrues. Thus, a neighbor adjacent to a collector could erect a structure that shaded the collector five, ten, or even twenty years after the collector was installed. If an obstruction were erected and then removed, the entire waiting period would have to begin again. (If the prescriptive period were set by legislation, as it is in England, the statute might also follow the British in allowing neighbors to file "light obstruction notices" that block the acquisition of rights to light without the erection of actual obstructions.[6])

American courts rejected the ancient lights doctrine on public policy grounds. A 1957 Florida case, *Fontainebleau Hotel Corp. v. Forty-Five Twenty-Five, Inc.*,[7] is typical. Although its holding applies only in Florida, it would have argumentative value in other jurisdictions. The Eden Roc Hotel sued the rival Fontainebleau when the latter proposed to build a fourteen-story tower that would shade the cabana, swimming pool, and sunbathing areas of the Eden Roc from two o'clock on during the busy winter months. Sunshine is the major reason tourists visit Miami Beach. Yet the District Court of Appeals of Florida supported the right of the Fontainebleau to shade the Eden Roc property: "No American decision has been cited . . . in which it has been held that—in the absence of some contractual or

4. Rights of Light Act, 1959, 7 & 8 Eliz. 2, c. 56. *See generally* Vincent Powell-Smith, *Defeating Acquisition of Right to Light,* NEW LAW JOURNAL 118 (1968): 855–56; H. W. Wilkinson, *Let There Be More Light,* ibid., pp. 7–8.

5. The history of this doctrine is discussed in WILLIAM A. THOMAS, ALAN S. MILLER, AND RICHARD L. ROBBINS, OVERCOMING LEGAL UNCERTAINTIES ABOUT USE OF SOLAR ENERGY SYSTEMS (Chicago: American Bar Foundation, 1978), p. 23; and David Bersohn, *Securing Insolation Rights: Ancient Lights, Nuisance, or Zoning?* COLORADO JOURNAL OF ENVIRONMENTAL LAW 3 (1976): 112–52.

6. Rights of Light Act, 1959, 7 & 8 Eliz. 2, c. 56, § 2.

7. 181 Fla. Supp. 74, 114 So. 2d 357 (Dist. Ct. App. 1959).

statutory obligation—a landowner has a legal right to the free flow of light and air across the adjoining land of his neighbor."[8] The court noted that under old English common law a landowner had a right to unobstructed light and air from over adjoining land after he had used that light, without interruption, for many years. However, in the United States, the opinion noted, the ancient lights doctrine has been unanimously repudiated[9] because public policy favors full development of land.[10]

Since courts rejected this ancient lights approach on policy grounds, it is conceivable they could also revive it on the ground that public policy now favors rights to sunlight, because of the energy crisis and new applications of solar technology.[11] In many areas of the law, such as pollution control and tenants' rights, there has been a strong trend away from the formerly unfettered right to develop and use land and toward consideration for the needs of the public. It can be argued that solar access rights should be similarly recognized.[12]

But even if the ancient lights doctrine were judicially or legislatively revived, it would not adequately protect solar access because it does not ensure access to direct sunlight. It was not meant to apply to energy conversion devices like solar collectors but to the daylight (often indirect light) passing through windows. The rules that English courts have applied to determine whether too much light has been taken away seem to be rules of thumb more than rules of law. One test is whether there is enough light in the middle of the room to enable a person to read without grumbling; another is whether light is unobstructed for 45 degrees above the window; yet a third is whether occupation of the house has been rendered uncomfortable

8. 114 So. 2d at 359.

9. Ibid.

10. *See generally* Franklin Gevurtz, *Obstruction of Sunlight as a Private Nuisance,* California Law Review 65 (1977): 94–119. Even in England, not all of the public favor the doctrine. More than a century ago Chancellor Kent observed that the doctrine of ancient lights "is not relished in this country." James Kent, Commentaries on American Law, ed. O. W. Holmes, Jr., 12th ed., 4 vols. (Boston: Little, Brown and Co., 1873), 3: 448.

11. Evidence of a change in public policy is found, for example, in the Solar Heating and Cooling Demonstration Act of 1974, § 2, 42 U.S.C. § 5501 (Supp. V. 1975); and the Solar Energy Research, Development and Demonstration Act of 1974, § 2, 42 U.S.C. § 5551 (Supp. V 1975).

12. *See* Daniel P. Moskowitz, *Legal Access to Light: The Solar Energy Imperative,* Natural Resources Lawyer 9 (1976): 193.

by obstruction of light according to the ordinary notions of mankind.[13] These tests are too capricious for solar access protection. Another problem with the doctrine is that it would not protect access that will be needed for collectors installed in the future.

Conclusions. The most serious disadvantage to the application of any prescriptive easement approach, including the ancient lights doctrine, is that a long prescriptive period must run between the first use of the light and the vesting of a right to continue to receive it. This waiting period before a right arises is what distinguishes prescriptive easements for light from other approaches, so it is not a fault that could be cured in a legislative version.[14] Another disadvantage is that the vesting of a prescriptive easement results from the accident of ownership over time, and not from thoughtful land use policy decisions. The uncertainty over whether a right would be perfected, coupled with the probability that neighbors would act to keep their property free of easements, is a strong reason for rejecting prescriptive easements as a realistic way to protect solar access.

Instant Access Rights

Legislation has been proposed on both the federal and state levels that would give solar energy users instantaneous, automatic rights to continued access to sunlight upon installation of collectors.[15] In one sense, these proposed laws are the antithesis of prescriptive easements, since they delete the need for time to pass before a right exists. But there are similarities. Like prescriptive easements, it is the accident of timing which determines rights, rather than rational planning by a government. And the extent of the right

13. W. A. Greene, *Right to Light,* SOLICITOR'S JOURNAL 112 (1968): 576.

14. If the period were eliminated, the result would not be prescriptive easements, but the "instant access" described in the next section of this chapter.

15. This discussion refers only to the federal bills, S. 985, 95th Cong., 1st Sess. § 2(a) (1977), and H.R. 259, 95th Cong., 1st Sess. § 2(a) (1977); New Mexico's Solar Rights Act, 1977 N.M. Laws ch. 169; and similar bills. It does not apply to legislation that significantly limits the scope of the solar rights that can be acquired or that attempts to coordinate the acquisition of rights with land use planning. The more limitations or standards are applied to instant access rights and the more factors like north slopes, authorized but unused high densities, and other troublesome situations are allowed for, the fewer problems there will be. Nonetheless, without case-by-case hearings to avoid undue hardships on neighbors (as in our prototype recordation ordinance), instant access rights will inevitably result in the kinds of problems discussed in the text in some cases. (And, if a hearing is provided for, the rights will no longer be instant.)

would depend on the extent of the use of sunlight, as it does in prescriptive easements.

At first glance, these "instant access" laws (which typically are very short) seem appealingly direct and simple: they give rights to sunlight that is needed, only when it is needed. Instant access approaches are sometimes advocated for two reasons. First, they seem easier to understand than more complex approaches like solar envelopes; their consequences, however, are really harder to understand. (See Chapter 2 and Appendix A.) Second, such laws are expected to encourage use of solar energy collectors by shifting the burden to affected neighbors. Neither of these qualities counterbalances the laws' disadvantages. In actuality, this approach would create many legal and political problems.

Although there is no typical language in the instant access bills, the bill introduced in Congress is typical in its sweeping effect:

> No State or local zoning law, regulation, ordinance, or other provision may permit the construction of any building or other object within the jurisdiction of such State or locality in any location or manner which would obstruct or otherwise interfere with sunlight necessary for the operation of any solar heating equipment, solar cooling equipment, or combined solar heating and cooling equipment which is in use on any building on the date on which any permit or other authorization for such construction is issued (or on the date of such construction in any case in which no such permit or other authorization for such construction is required under the applicable law, regulation, ordinance, or other provision).[16]

New Mexico enacted a law with a similar effect in 1977 (see Appendix B).[17] And bills resembling New Mexico's law have been introduced in at least four states.[18] Under these bills, a landowner would receive a valuable benefit by installing a solar energy collector while severely restricting development of adjacent land to the east, south, west (and somewhat to the north for cooling or hot water collectors). In fact, under most of these proposed laws, if a solar doghouse with a collector on its south wall were placed on a southern lot line, it might prevent nearly all further development on the lot to the south.

16. S. 985, 95th Cong., 1st Sess. § 2(a) (1977), and H.R. 259, 95th Cong., 1st Sess. § 2(a) (1977). As discussed in Chapter 12, such federal solutions do not appear appropriate to this particular land-use issue. *See also* Brown v. Environmental Protection Agency, 521 F.2d 827 (9th Cir. 1975), in which the court declined to validate an implementation plan under the Clean Air Act prepared for California by the federal government because it imposed affirmative duties on the citizens of that state.

17. 1977 N.M. Laws ch. 169 (reproduced in Appendix B).

18. These include R.I. S. 2341 (1978); Okla. H.B. 1741 (1978); N.Y. A. 1706 (1977) and A. 607 (1977); and Mo. H.B. 715 (1977) and H.B. 1216 (1977).

To simplify the discussion of this approach, it is assumed that the statutory language is sufficiently specific not to raise enforceability problems due to vagueness.[19] Further, in the case of the federal bills, it is assumed that Congress could legitimately stretch its powers to regulate interstate commerce and provide for the national defense to the protection of solar access.[20]

Unlike most other proposed methods of protecting solar access, the instant access laws described here could so greatly restrict the development of some nearby land that no reasonable use of that would be left. For example, a solar energy system that saved the owner of one lot $100 a year in fuel could prevent the construction of a $5 million building on another lot. An unconstitutional taking of property could thus occur (see Chapter 3). Owners might feel compelled to develop their land prematurely, before a neighbor installed a collector, just to protect their property rights. Conceivably, a person might install a collector only to control development on adjacent lots; neighbors would then have to buy their development rights from him. The fact that development rights may change over time will also affect property values. This uncertainty may make investors reluctant to buy land for future development. As nearly identical lots in the same neighborhood could be subjected to substantially unequal restrictions or could receive unequal solar access benefits, equal protection problems would arise.

Conclusions. A federal, state, or local "instant access rights" rule for allocating solar energy is undesirable for constitutional and policy reasons. The restrictive effect of an unqualified solar right on the development of neighboring property would often be unacceptable, not only for the land directly affected but also for comprehensive land use planning. Far less intrusive alternatives are available.

Water Law

Natural resources law is replete with analogies. Judges, legislators and commentators frequently propose analogies between natural resources in quests for legal provisions to govern the allocation

19. Although "vagueness" questions arise more commonly in criminal actions, the unsettling consequences of vague language extend beyond those cases.

20. Although Congress has very broad powers in these areas, it is doubtful they should be exercised at the present time.

of a new resource.[21] Commentators have proposed such analogies between sunlight and other natural resources. Water law[22] was suggested as a useful guide to solar access rights in 1975;[23] and in 1976 a law review comment[24] proposed state legislation declaring that the right to use solar energy is a property right and specifying that water law is applicable to disputes over the use of solar energy where practical.[25]

Judicial Implementation

The author of the comment suggested that a court dealing with a solar access case would have no applicable statute or cases, so that it could develop a new approach or use the principles for an analogous natural resource. According to the author, since "allocation problems for sunlight are similar to those for water resources the judiciary should seriously consider the adoption of water law principles for sunlight."[26] Contrary to the assertion in the comment, courts do have centuries of precedents to decide a case concerning

21. In retrospect, some of the analogies seem forced. The Kentucky Supreme Court heard arguments 125 years ago in which the plaintiffs likened the oil in question to solid minerals and the defendants suggested the analogies of animals *ferae naturae* and waters. Hail v. Reed, 54 Ky. 479 (1854). Perhaps the case most noted for analogies is Hammonds v. Central Ky. Natural Gas Co., 255 Ky. 685, 75 S.W.2d 204 (1934), in which the Kentucky court ruled that gas that had been stored in subterranean chambers had reverted to nature and therefore could be withdrawn through oil wells belonging to surface owners. The court also relied upon an analogy with percolating water, and then quoted an authority on oil and gas for the proposition that removing oil from the earth and then returning it could be analogized to severing timber from the soil and then converting it into realty again by using the lumber to construct a building. The holding in *Hammonds* has been thoroughly repudiated, either by statute or case law, as in White v. New York Natural Gas Corp., 190 F. Supp. 343 (W.D. Pa. 1960).

22. Two basic approaches to water law are used in the United States: the prior appropriation doctrine favored in the West, and the riparian doctrine of the East. The description and analysis of water law in the text refers to the western version. Under riparian law, each owner of land bordering a stream typically has equal rights to the water in the stream. Riparian states may either say that owners have a right to an undiminished "natural flow" or to not be subject to "unreasonable uses" by others. However, the trend in riparian states is toward permit systems.

23. WILLIAM A. THOMAS, ED., PROCEEDINGS OF THE WORKSHOP ON SOLAR ENERGY AND THE LAW (Chicago: American Bar Foundation, 1975), p. 10 (comments of Charles Haar).

24. Mary D. White, *The Allocation of Sunlight: Solar Rights and the Prior Appropriation Doctrine,* UNIVERSITY OF COLORADO LAW REVIEW 47 (1976): 421–47.

25. Ibid., p. 444.

26. Ibid., p. 447.

allocation of sunlight. The ancient maxim *Cujus est solum est usque ad coelum* ("the owner of the soil owns also the sky") still summarizes, with some qualifications, the right of a landowner to occupy the space above the surface of land to the height needed for reasonable use and enjoyment of the property.[27] The *Fontainebleau* case, discussed earlier, is only one of many cases that validate this element of U.S. property law in relation to sunlight. Courts are unlikely to depart from settled property doctrine to redistribute rights on the basis of tenuous water law analogies: a court might be violating established restraints on judicial activism if it did so, and the holding could be unconstitutional for that reason alone.

Legislative Implementation

Whatever the merits of judicial action to assign rights to solar energy in conformity with rights to water, courts are not expected to adopt the analogy. State legislators could, however, follow New Mexico's example: the New Mexico Solar Rights Act[28] is similar to the "model act" suggested in the law review comment. Under the New Mexico statute, prior appropriation requires that:

> In disputes involving solar rights, priority in time shall have the better right except that the state and its political subdivisions may legislate or ordain that a solar collector user has a solar right even though a structure or building located on neighborhood [sic] property blocks the sunshine from the proposed solar collector site.[29]

A "solar right" is defined as "a right to an unobstructed line-of-sight path from a solar collector to the sun, which permits radiation from the sun to impinge directly on the solar collector."[30] The

27. For historical analyses of rights to receive sunlight, *see*, e.g., ENVIRONMENTAL LAW INSTITUTE, LEGAL BARRIERS TO SOLAR HEATING AND COOLING OF BUILDINGS (Springfield, Va.: National Technical Information Service (HCP/M2528-01), 1978), pp. 4–6; and THOMAS, MILLER, AND ROBBINS, OVERCOMING LEGAL UNCERTAINTIES ABOUT USE OF SOLAR ENERGY SYSTEMS.

28. 1977 N.M. Laws ch. 169. The act in full is reproduced in Appendix B. The law has been criticized for the following problems, which are unrelated to its basis in water law: (1) insufficient specification of easement size; (2) no distinction between high- and low-density areas; (3) no reference to vegetation; (4) no reference to land slope; (5) poor guidance to political subdivisions on how to establish and administer regulations; and (6) lack of tradeoff flexibility for technically and economically balancing solar access needs and easements burdens. Robert F. Wells, *New Mexico Solar Access Law: Problems with Implementation of the Law,* SOUTHWEST BULLETIN 11 (Santa Fe: New Mexico Solar Energy Association, 1978), p. 9. *See also* Karin H. Hillhouse, *Problems and Prospects for the Pacesetter,* ibid., pp. 14–17.

29. 1977 N.M. Laws ch. 169, § 4(B)(2).

30. Ibid., § 3(B).

statute also provides that solar rights are freely transferable in accordance with normal recording procedures.[31] Another provisions allows county and municipal zoning authorities to have "permit systems" for the use and application of solar energy," unless "singular overriding concerns occur which significantly affect the health and welfare of the citizens of [the] state."[32]

Uncertainties created by water law doctrine. Under the prior appropriation doctrine, a person obtains a vested right to use a particular amount of water by diverting it from a navigable watercourse and applying it to a beneficial use within a reasonable time.[33] The user then has a right to continue to divert that amount of water, which precludes upstream users from using so much water that this prior appropriation cannot be fulfilled.[34] But what constitutes a "beneficial use" of solar energy under the New Mexico act is unclear. Beneficial use under the act depends on use of a "solar collector," which is

> . . . any device or combination of devices or elements which rely upon sunshine as an energy source, and which are capable of collecting not less than twenty-five thousand Btu's on a clear winter solstice day. The term also includes any substance or device which collects solar energy for use in:
> (1) the heating or cooling of a structure or building;
> (2) the heating or pumping of water;
> (3) industrial, commercial or agricultural processes; or
> (4) the generation of electricity.
> A solar collector may be used for purposes in addition to the collection of solar energy. These uses include, but are not limited to, serving as a structural member or part of a roof of a building or structure and serving as a window or wall. . . .[35]

The 25,000 Btu standard would include the southfacing windows on most buildings that have the glazing required under building codes. (Such glazing is used by custom even where not required.)

31. Ibid., § 4(B)(3).

32. Ibid., § 4(C).

34. *See,* e.g., State v. Miranda, 83 N.M. 443, 493 P.2d 409 (1972).

34. The doctrine evolved during the early development of the arid West, where three uses—mining, irrigation, and stock watering—accounted for nearly all of the water appropriated. The West was then sparsely populated, and many natural resources were available to be claimed. With increased development of our western states, the doctrine was modified by comprehensive state water codes into an administrative process that generally retains the elements of the doctrine but adds requirements for permits, public hearings, and other procedures. Wells Aleck Hutchins, Water Rights Law in the Nineteen Western States (Washington, D.C.: U.S. Government Printing Office, 1977), chaps. 5–9.

35. 1977 N.M. Laws ch. 169, § 3(A).

But the "also includes" language and item (1) would remove the need for supplying even this Btu contribution; most existing buildings in New Mexico would appear to have solar collectors protected under this act. Thus, urban development in the state could be virtually stopped in the unlikely event this statute were found constitutional. Even if the act's terms were narrowly defined, water law provides no basis for granting a preference to one type of beneficial use over another. Under water law, priority in time determines which beneficial use prevails. If a beneficial use concept is to apply in solar access law, it would be better to limit it to socially advantageous uses. For instance, it would be foolish to allow high density construction near major public transportation routes to be precluded by the heating of a solar doghouse, the growing of tomatoes four months a year, or by the drying of crops one month a year.

In practice, the right under the prior appropriation doctrine commonly relates to the date an application to divert water is filed with the appropriate state agency, rather than the date when all elements of the doctrine are met. The author of the law review comment suggests that the time of vesting should relate back to the time construction begins.[36] Landowners and lenders are entitled to something more reliable. Construction itself might provide some notice to adjacent property owners, but this notice would be ineffective because the right already would have been vested. How would property owners know whether, for example, a window in a building under construction on adjacent property would bar future construction on their own? Constant surveillance of surrounding property would be necessary. Under a permit system that dates the right from the time of application, landowners would be forced to monitor permit applications instead of construction activity. Public hearings or other means of providing notice would also be ineffective if the right springs to life upon filing.

The "beneficial use" criterion has another problem. The more hours of the day collectors use solar energy, the more hours of protection the collector owners are entitled to[37] and the more they burden nearby lands. (There may be some reasonable limits on this from water law, but the limits for solar energy are unclear.) A small or inefficient system might beneficially use sunlight from dawn to dusk, whereas a large or efficient system could require (and thus be

36. White, *Allocation of Sunlight,* pp. 437–38.

37. 1977 N.M. Laws ch. 169, § 4(B)(1).

protected) fewer hours. System operating characteristics, use patterns, and anything else that affects loads on the solar system would affect the hours of sunlight needed and thus the restraints on development of nearby properties.

To counterbalance this problem, however, the collector owner would have an almost impossible burden of showing the actual hours of beneficial use that the system has had over the course of its use. A recording mechanism to keep track of hours of operation would be possible for active systems (using industrial pen recorders, for example); for passive systems no means of recording is apparent. Even if it suffered no other disadvantages, we would doubt the feasibility of the "beneficial use" approach, both because it could impose unreasonable restraints on development and because it would require impractical gathering of evidence by the solar user to prove entitlement to the solar rights.

Conclusions. Water law would be a poor way to protect solar access. It suffers from all the defects of other "first-come-first-served" approaches, raises constitutional issues of equal protection and due process, and could not protect access to sunlight that will be needed in the future. A statute more carefully written than the New Mexico act could avoid some of the problems with a water law analogy. One possibility would be to prune the rights granted to a reasonable size. New Mexico's law provides all the protection collectors of any size in any location can "beneficially use." Another possible refinement would be to specify what uses would prevail in certain situations where a proposed structure would shade a collector. However, such statutory adjustments to water law would be cosmetic only. Because water law is totally unrelated to the land use planning process, arbitrary and discriminatory treatment of property would probably result anyway. If an attempt were made to integrate a water law type statute with a comprehensive plan, the distinguishing hallmarks of water law would be so eroded that it would be preferable to start over with a completely new law. A well-written law need not refer to water law. The usefulness of analogies is not in question, but there is no need to adopt an entire body of law simply to benefit from some of its features.

Part IV
Compensatory Approaches

Most of the solar access protection ideas discussed thus far have relied on the police power and have not, therefore, involved compensation to the owners of restricted properties. As a practical matter, noncompensatory methods of protecting solar access are preferable.

Exceptions to this general rule do occur. One is where an individual wishes more access protection than a law grants to everyone in his situation. As this additional access will impose additional restrictions on his neighbor's property, it is only fair that the neighbor be compensated. Ideally, he and his neighbor will be able to agree privately on an appropriate amount of compensation for a solar easement. Private agreements are the subject of Chapter 10. They may take the form of restrictive covenants as well as easements and will be useful in many situations.

Compensation from public coffers may also sometimes be desirable. This compensation could be in either cash or transferable development rights. If a large public building using south-wall collectors were erected, for example, eminent domain could properly be used to condemn solar easements over lots immediately to the south. Other possible roles for eminent domain are explored in Chapter 11. The more unusual compensatory approaches of transferable development rights and zoning by special assessment financed eminent domain are also evaluated as possible tools for protecting solar access.

Although the law has set limits on how public monies may be spent, the constraints on compensatory approaches are really more economic than legal. As attorney Nicholas C. Yost has said:

> When it comes to the compensation issue, I am sure nobody would object to compensating everybody for everything, except that the money just isn't there. We're dealing with the very real problem of budget constraints which are determined by the amount of public funds it is possible to acquire.[1]

1. PRIVATE PROPERTY AND THE PUBLIC INTEREST: SUMMARY OF PROCEEDINGS OF A CALIFORNIA TOMORROW LEGAL SEMINAR ON THE USE AND REGULATION OF LAND (San Francisco: California Tomorrow, February 1974), p. 14.

10 Private Agreements

Where there is no public solar access protection, private arrangements present an attractive option;[1] even where there is public solar access protection, they can be a useful supplement. Properly drafted private agreements are legally enforceable, offer protection with a minimum of government involvement, and permit a large degree of site-specific control over the protection afforded.

Private arrangements can take the form of restrictive covenants regulating entire subdivisions or of individual agreements (creating easements, real covenants, and equitable servitudes) that bind two neighbors. The law recognizes both types of arrangements; if an agreement is broken, the injured party has a judicial remedy.

Restrictive Covenants

The most common private controls on land development and use are restrictive covenants.[2] Typically, the developer incorporates these covenants in the deed to each lot sold in a subdivision. It is also possible for a group of neighbors to agree to restrictions on the use of their property and have the covenant recorded so as to run with

1. *See* American Planning Association, Protecting Solar Access for Residential Developments: A Guidebook for Planning Officials (Washington, D.C.: U.S. Government Printing Office, 1979), for further discussion of private agreements to protect solar access.

2. *See generally* Norman Williams, Jr., American Land Planning Law, 5 vols. (Chicago: Callaghan & Co., 1975), 5: chap. 154, pp. 241–53.

the titles to each lot. Any neighbors who did not sign the agreement would not be bound, however.[3]

To protect solar access in new subdivisions, a developer can insert a restrictive covenant in each deed that prohibits structures or vegetation above a certain height on specified areas of any lot. A restrictive covenant usually binds subsequent owners of the land, but to do so it must meet certain technical requirements:[4]

- □ The purchaser always must have notice of the obligations; this is satisfied by recording in the public land records.
- □ The person seeking enforcement usually must own some land protected by the covenant and sometimes must have "privity of estate."
- □ The obligations of the covenants sometimes must "touch and concern" land.
- □ The obligations of the covenant usually must be negative (prohibiting a use or activity) rather than affirmative (requiring a landowner to take action).

Covenants that limit the placement and height of structures and vegetation to protect access to sunlight could generally meet all these requirements. The obligation not to let vegetation grow above a certain height might cause a problem, however, because the owner must trim the vegetation. This might be construed by some courts as an unenforceable affirmative action. A partial solution might be to reword the covenant so that it prohibited the planting, in particular places, of species of trees or bushes that could grow to over a specified height.

Courts generally construe restrictive covenants narrowly and resolve ambiguities in favor of the unrestricted use of land. To avoid

3. Sometimes covenants are used instead of zoning. For example, Houston has no zoning ordinance, but nearly all the residential areas are subject to restrictive covenants. BERNARD H. SIEGAN, LAND USE WITHOUT ZONING (Lexington, Mass.: Lexington Books, 1972), p. 78. Houston also uses subdivision controls (*see* Chapter 6, note 2, and associated text). Covenants are imposed at the time of platting and contain controls such as lot size and setback requirements typically found in zoning ordinances. The city of Houston can enforce the covenants to the same extent as a property owner. TEX. REV. CIV. STAT. ANN. art. 974a-1 (Vernon Supp. 1970). Houston seems to be the only major U.S. city using this method of controlling land use.

4. For a more detailed discussion, *see* WILLIAMS, AMERICAN LAND PLANNING LAW, 5: §§ 154.04–.08, pp. 244–48.

this result, the terms of the covenant should be defined clearly and its purpose should be described in detail.

It will be exceedingly difficult to establish restrictive covenants to protect solar access in areas that are already developed. Convincing many neighbors to agree to be bound by a covenant will seldom be simple. Collector owners will probably have to use solar easements.

Solar Easements

Another alternative to public regulation is the use of easements created by private agreements between two (or a few) persons to protect the right of one to solar access. These easements can produce the same results as restrictive covenants, only on a smaller scale. (Restrictive covenants convey the same rights as reciprocal negative easements among all the parties, so that many are benefited as well as bound.) It is usually accepted that a person can bargain for an express easement to light, air, and view.

A landowner receiving the benefit of the easement may pay a flat fee to purchase the easement or may lease the easement from the other landowner;[5] or the parties may make reciprocal agreements not to block each other's sunlight. Agreements between neighbors have several advantages over public remedies: they may be shaped to fit individual site and system requirements; they involve no public expenditures, red tape, or permission; and they are more flexible than restrictive covenants.

Privately negotiated solar easements are, however, far from being an adequate mechanism for guaranteeing solar access:

- Easements are voluntary. A neighbor cannot be forced to sell one.
- Although it may be easier to negotiate easements before conflicts arise, few collector owners might do so.
- The costs of purchasing a solar easement may reduce the cost-effectiveness of solar energy use.[6]

5. In Macht v. Dept. of Assessment of Baltimore, for example, a lease of an easement for light and air was held valid and the property tax of the lessor was increased to include the value of the easement. 266 Md. 602, 296 A.2d 162 (1972).

6. No data are yet available on the price of solar easements.

- Transaction costs may also be great, especially if the collector owner required legal and technical advice.[7]
- The purchase of an easement might give a windfall to the owner of "burdened" land who never really intended to use his land in a way that would block sunlight.
- A solar collector owner would often have to include several neighboring property owners in easement negotiations if solar access is to be protected adequately.[8] As the number of people increases, the probability of reaching an economically feasible arrangement diminishes.

Although private agreements to create airspace easements are legal everywhere in the United States, many states have enacted or are considering legislation that would explicitly recognize solar easements.[9] Such statutes can:

- Make it clear that private agreements will bind and benefit future owners of the parcels.
- Standardize the form for recording solar easements, simplify their implementation, and reduce the need for costly legal drafting.
- Create public awareness of solar easements and encourage their use.

7. In simple situations (typical installations on level sites), these costs might be avoided by using model easement forms prepared by solar experts, perhaps provided by the local or state government. Comparable forms are available for separations, divorces, wills, etc. If these were properly designed for the locality in question and accompanied by appropriate explanations of when professional advice would be wise, they could cut costs without much risk to the parties.

8. This would include, at a minimum, neighbors to the south, southeast, and southwest. It would probably include those to the east and west as well, and possibly even those to the northeast and northwest (depending on the system). More persons would have to be included in high-rise areas than in low-rise. (See Chapter 2 and Appendix A).

9. In 1975, Colorado enacted a law to facilitate private solar agreements (1975 Colo. Sess. Laws ch. 326). Similar statutes have been passed in California (A.B. 3250, passed 31 August 1978), Florida (1978 Fla. Laws ch. 309), Georgia (1978 Ga. Laws, Act 1446), Illinois (1978 Ill. Laws, Pub. Act. No. 80-430), Kansas (1977 Kan. Sess. Laws ch. 227), Maryland (1977 Md. Laws ch. 934), Minnesota (1978 Minn. Laws ch. 786), North Dakota (1977 N.D. Sess. Laws ch. 425), and Virginia (1978 Va. Acts ch. 323). Other states considering similar legislation include Arizona, Hawaii, Missouri, New Jersey, Ohio, Pennsylvania, Rhode Island, Tennessee, Utah, and West Virginia.

- □ Be politically acceptable because they cost states almost nothing.

Most proposed or enacted solar easement laws have two basic provisions. First, solar easements must be in writing and are subject to the same conveyancing and recording requirements as other easements. Second, certain specific terms must be included in a solar easement, including (1) the angles at which the solar easement extends over the restricted property, (2) any terms or conditions that specify the length of the grant or make provisions for its termination, (3) any provisions for compensating the owner of the property benefiting from the solar easement in the event of interference with the use of the easement, and (4) any provisions for compensation of the owner of the property burdened by the solar easement for maintenance of the easement. While there is no harm to items (2) through (4), there is no clear need to rule out easements in terms of time instead of space as required by (1). Although most laws and proposals do not rule out easements in terms of time, by implication they at least raise serious doubts over the validity of easements expressed in time rather than degrees. (See Chapter 2 and Appendix A for discussion of relative merits of these two approaches.)

Another limitation in most laws and proposals is that solar access easements are authorized for protecting access for solar collectors.[10] The question may thus be raised whether an easement for solar access would be valid and recordable if it was for a solar collector that did not fit the state's legislative definition of a collector, or if an easement is sought for a site where a collector may be installed in the future. Such solar access easements should not be excluded, even by implication.

A statute along the following lines should be sufficient for authorizing solar access easements:[11]

> "Solar access easement": An easement that limits the heights, locations, or both, of permissible development on the burdened land in terms of structures, vegetation, or both, for the purpose of providing access for the benefited land to direct solar energy passing over the burdened land.

10. For example, " 'Solar easement' means the right of receiving sunlight across real property of another *for any solar energy system*" (emphasis added). Cal. A.B. 3250 (passed 31 August 1978). New York's S. 8561-A, A. 11384-A (1978), is an exception.

11. New York S. 8561-A, A. 11384-A, § 2 (1978) (to be codified as § 13-105 of N.Y. Energy Law), is even more abbreviated. It does not define the easement. A definition may not be needed.

A solar access easement shall be in writing and shall be subject to the same conveyancing and recording requirements as are other easements. Solar access easements shall run with the lands benefited and burdened unless otherwise expressly stated therein.

Conclusion

Restrictive covenants and easements are two forms of privately negotiated arrangements that can help protect solar access. Restrictive covenants can be particularly useful in new developments; easements can be helpful in existing developments. Nevertheless, private arrangements offer only a partial solution and are not a substitute for public solar access laws.

11 Eminent Domain, TDR, and ZSAFED

The use of eminent domain to protect solar access may be considered a last resort measure by a local government. As discussed in Chapter 3, certain public use and public necessity requirements must be met before the eminent domain power may be exercised. However, just compensation and procedural requirements must also be met. For solar access protection, a locality may wish to explore two alternatives to the payment of public money for the condemned property: transferable development rights (TDR) and zoning by special assessment financed eminent domain (ZSAFED).

Eminent Domain

Eminent domain is a longstanding and well-established power of a sovereign government to take private property for public use or public benefit without an owner's consent, subject to the payment of just compensation. This power has been used to condemn highway and railroad right-of-ways and scenic easements; to clear land for urban renewal; to provide easements for electrical, sewage, and other utility lines; and on occasion, to preserve historic districts by condemning potential development rights. Eminent domain has proven to be an effective government tool. It represents the strongest form of public land use control in that the government actually acquires (rather than merely regulates) a private property interest.

This tool could be the last resort for public protection of access to the sun. By using eminent domain to acquire a solar access easement from a property owner, a local government would, in effect, be paying the property owner to prevent him from developing in a manner that would shade his neighbor's solar collector. If eminent domain is successfully pursued, a municipality could avoid any complaints of unfair treatment or of unconstitutional deprivation of property on the part of the landowner.

Just Compensation Requirements

To protect solar access with eminent domain, local governments would not acquire a right to make use of property but would take an easement that would prevent a property owner from obstructing his neighbor's solar access through either construction or excessive vegetation growth. This taking would have to be justly compensated.

While there are several methods for calculating just compensation for an easement, the method used most often evaluates the property both before and after the easement is taken and sets the compensation at the difference between the two values. The difference represents the value of the easement plus severance damages. Severance damages are the diminution in the value of the remaining property caused by the loss of the taken property.[1] For vegetation, such damages would occur when the trimming or removal of existing trees and shrubs to prevent shading of a nearby solar collector seriously diminished the value of the entire property. Compensation for an easement may include the diminution in value of the trimmed vegetation, trimming costs, and compensation for loss of shading (which could affect cooling loads). Compensation generally would not cover the aesthetic value of vegetation, itself, since few states consider aesthetics, *per se*, in determining loss in value, but any diminution of the market value of the property taken as a whole due to the change in vegetation aesthetics would be considered.[2]

When an easement is sought over a structure, the value of the easement depends upon the amount of space above the existing structure that could economically be developed. In predominately single-family residential areas such easements would probably be of negligible value. If, under existing zoning ordinances, the landowner has no right to develop above the existing height of his home, a negative easement above his home would have no value. In an area zoned to allow development beyond existing heights, especially if the

1. JULIUS L. SACKMAN, NICHOLS' THE LAW OF EMINENT DOMAIN, rev. 3d ed., 14 vols. (New York: Matthew Bender, 1976), 4A: § 14.1[3], pp. 14-31–35.

2. *See* ibid., 4A: § 14.243, pp. 14-175–177; 4: § 13.21, pp. 13-86–95; 5: § 16.1013[4], pp. 16-53–59, esp. at n. 43, p. 16-58; and, 5: § 18.41[2], n. 43, p. 18-153.

property is zoned for skyscrapers, multi-story office buldings, or hotels, the easement could be valuable.

There is no obvious source of funds to pay for condemned solar easements. Benefited solar collector owners could be assessed for the value of the easement. General revenues could be used, although funding for other purposes would probably be reduced. Local legislators would be reluctant to dedicate funds to condemnation of solar access easements before a strong public interest in taking advantage of solar access can be demonstrated. At that point, however, such funding may be justified. General obligation bonds could be used if the issuance would benefit the city as a whole. Assessment bonds must also satisfy a public purpose test, but, in addition, they must give a special benefit to the individual properties that are assessed. The question is whether voters would be willing to approve a bond issue for solar access, especially solar access through use of eminent domain. Voters often react unfavorably to eminent domain, whatever the purpose.

Procedural Requirements

In addition to the compensation problems involved in eminent domain, arduous procedural hurdles must be overcome. Because of the federal Constitution's due process requirements, condemnation proceedings are characterized by complex, lengthy administrative processes and other complicated procedural requirements: In judicial condemnations, notice and opportunity to be heard must precede any determination of a compensation award; in administrative condemnations, notice and an opportunity for a hearing before an impartial tribunal is granted only after the award has been fixed. The rules concerning who may qualify as an appraiser to determine compensation vary from state to state. Most states require that clear and detailed compensation reports be made so that an aggrieved property owner can seek judicial appeal. If individuals who are benefited by the use of eminent domain are then assessed for the value of whatever they received, a second appraisal is required.

When Eminent Domain May Be Useful

Eminent domain is often confused with takings, although they are really distinct issues with different roots and consequences. A taking is said to occur when a government has used its police power to regulate land but the regulation is so stringent that no reasonably beneficial use of the land remains. Under a taking, an interest in land does not shift to the government as it does when eminent domain powers are used. When there is a taking, the injured party may seek either of two remedies: to have the action taking his property declared void or to be compensated for the taking (known as inverse condemnation). (See Chapter 3 for a discussion of takings.)

Eminent domain could be used where lawsuits arguing taking or lack of a valid public purpose are likely: for example, in areas where only a few properties are suitable for solar retrofit, police power regulations such as zoning might be successfully challenged by landowners who would not benefit from the regulation but who would have to bear the burden of it. Instead of zoning, a municipality could condemn solar access easements, compensate restricted landowners for them, and then assess collector owners for the value of the easements. This approach could be costly but would provide more security for the solar investor. The main legal obstacle to this approach is that if only a few persons benefit from solar easements, courts may not find a strong enough public purpose to justify the use of eminent domain.

A recent trend in land use cases may lessen the need to use eminent domain for solar access protection. Some state courts have maintained that the right to develop property is no longer an inherent part of property ownership. Under this interpretation, the "condemnation" of a solar access easement above someone's property would not constitute a taking because the landowner has no inherent right to use the condemned space. This trend has come about partly because of new and more flexible zoning techniques, but more directly as a result of expanding notions of public rights.[3]

As public intervention under the police power has become more acceptable, use of eminent domain has decreased. In situations other than highway, railroad, and utility easements, or urban renewal, eminent domain powers are seldom used. The reasons for this are the complex and time-consuming administrative and procedural requirements, costs of litigation, and costs of compensation to restricted landowners.

Transferable Development Rights

Despite its novelty, the transferable development rights device (TDR) has been the subject of a burgeoning bibliography of aca-

3. *See generally* Sondra E. Berchin, *Regulation of Land Use: From Magna Carta to a* Just *Formulation,* UCLA Law Review 23 (1976): 904–35.

demic literature and has been reviewed by state and federal legislative study commissions.[4] Interest in TDR has been increased by the *Penn Central* decision, which contains language encouraging to TDR advocates.[5] It is understandable, therefore, that investigators wrestling with the stubborn issues attending the protection of solar access would wish to consider TDR.

It is possible to explore, in a preliminary way, TDR's major advantages in preserving solar access.[6] ELI suggests that state enabling legislation allow a locality to experiment with TDR for solar access protection even though TDR is not expected to be widely used for this purpose.

An Overview

Neither fish nor fowl, TDR schemes are sometimes based on eminent domain powers and sometimes on police powers. Basically, TDR distinguishes between the right of ownership and the right to develop: it allows all or part of the right to develop to be sold and to be applied (transferred) to other properties. The quantum of development rights for any lot is that permitted by current zoning but not used. So portrayed, TDR seems simple enough. In fact, there are many variations on this basic pattern: the owner of the underdeveloped parcel may be permitted to sell his rights only to a public agency that simply retires them,[7] or that banks them for subsequent sale to other private landowners.[8] Or the owner might be given the

4. *See* J. V. Helb, B. B. Chavooshian, and G. H. Nieswand, eds., Development Rights Bibliography, New Jersey Cooperative Extension Service Leaflet no. 53 (New Brunswick, New Jersey: Rutgers–The State University, Cook College, 1976), pp. 1–10, 21–30, 53–67.

5. Penn Central Transp. Co. v. City of New York, 46 U.S.L.W. 4856, 8 Envir. L. Rep. 20528 (1978). This case is discussed in Chapter 3. The Court refers to TDR in 46 U.S.L.W. at 4865; in 8 Envir. L. Rep. TDR are discussed at 20536.

6. For a somewhat different view *see* Jesse L. Matuson, *A Legislative Approach to Solar Access: Transferable Development Rights,* New England Law Review 13, no. 4 (1978): 835–76. A less optimistic assessment of TDR's utility in solar access protection is offered by Dale D. Goble, *Solar Rights: Guaranteeing a Place in the Sun,* Oregon Law Review 57, (1977): 127–28. Goble notes "The initial reception in the courts of the TDR concept has been lukewarm at best." Ibid., p. 128.

7. As in programs in New Jersey and in Suffolk County, New York. *See* Phillip Alampi, *The Blueprint Farmland Preservation Concept as a Tool,* New Jersey Municipalities 6 (Trenton, New Jersey: New Jersey State League of Municipalities, November 1975): 6–7, 30–32; and Klein, *Report of the Suffolk County Legislature from the Select Committee on the Acquisition of Farmlands* (Suffolk County, N.Y., Office of the County Executive, 1974).

8. John J. Costonis and Robert S. DeVoy, The Puerto Rico Plan: Environ-

options of selling the rights to another private landowner, of using the rights on some other lot that he owns, or of selling the rights to a public development rights bank.[9]

Types of land subject to TDR. If TDR's purpose is to protect landmark buildings, the lots from which transfer is allowed are those containing designated landmarks. For agricultural preservation programs, the lots will be those devoted to farming. If better urban design based on the density zoning principle is the object,[10] the lots will be those to be held in a low density status. For solar access purposes, the lots will be those across which solar access is to be protected, presumably because a solar collector is installed or is to be installed nearby.

Selecting lots or districts to which rights may be transferred is more difficult. Some TDR programs limit transfers to lots that are adjacent to or across the street from the transferred lot.[11] For solar access purposes, this often would not work. To avoid shading, the transferee lot would have to be farther north or away from the protected collector site (or even away from the area zoned for solar access). Other TDR programs enlarge the transferee locations to include entire areas selected on the basis of the capacity of their physical infrastructure, public services, and urban design characteristics to support the added development due to such rights as may be transferred to them.[12] This alternative would appear more feasible for solar access protection. The transferee district might also be one that has little or no solar potential. Such districts could include north slopes, areas overgrown with tall trees, or areas with such mixed

MENTAL PROTECTION THROUGH DEVELOPMENT RIGHTS TRANSFER (Washington, D.C.: Urban Land Institute, 1975).

9. *See* JOHN J. COSTONIS, SPACE ADRIFT: LANDMARK PRESERVATION AND THE MARKETPLACE (Urbana, Ill.: University of Illinois Press, 1974).

10. Under this principle, density is not allocated on a lot-by-lot basis, each lot subject to the same stated maximum; instead, it is allocated globally for an entire area, density allowances on individual lots being freed from a uniform maximum. *See generally* E. LOVELACE AND W. WEISMANTEL, DENSITY ZONING: ORGANIC ZONING FOR PLANNED RESIDENTIAL DEVELOPMENT, Urban Land Institute Technical Bulletin no. 42 (Washington, D.C., 1966); Eli Goldston and James H. Scheuer, *Zoning of Planned Residential Developments,* HARVARD LAW REVIEW 73 (1959): 241–67; and *Symposium: Planned Unit Development,* UNIVERSITY OF PENNSYLVANIA LAW REVIEW 114 (1965): 1–170.

11. Transfers permitted under the New York City landmarks program are severely restricted in this way. *See* John J. Costonis, *The Chicago Plan: Incentive Zoning and the Preservation of Urban Landmarks,* HARVARD LAW REVIEW 85 (1972): 572, 585–89.

12. *See* COSTONIS, SPACE ADRIFT, chap. 2.

heights that public solar access protection seems unmanageable. Complex planning and administrative questions are posed as the volume of transferrable rights and the size of the transferee area are increased.

Economic questions. There are two basic economic issues: first, what quantum of rights should be made transferrable? For solar access protection, this would be the amount necessary to avoid shading the protected collector. Second, will there be a market for the rights in the transferee area? This depends on the market demand for new construction there and the extent to which this demand cannot be satisfied under current land use controls, excluding the TDR scheme.[13]

TDR's purpose. Although TDR may serve a variety of goals, such as the preservation of particular structures or areas, its main purpose is to overcome what has been labeled the "windfall-wipeout" dilemma.[14] That is, some kinds of public land use regulation help some landowners while hurting others. (Solar access protection on a lot-by-lot basis would be an example.) In preventing significant further development or redevelopment on some lots, these regulations lessen the economic return on them, sometimes drastically and haphazardly. They may also increase the value of non-restricted lands, if there is a strong demand for development in the area.

TDR's adherents argue that it is valid not only because it advances independent public purposes (such as landmark preservation), but also because it does so in a manner that deals fairly with those benefited and burdened. Justice Brennan, in his *Penn Central* opinion, emphasized that TDR schemes mitigate the financial burden and are thus to be considered when determining the impact of restrictions on development.[15]

Application of TDR to Solar Access

Without a specific economic, planning, and engineering setting, it is difficult to generalize about TDR's usefulness to solar access protection. TDR might be applicable in the following situations:

13. Ibid., chap. 4

14. Hagman, Donald G. *A New Deal: Trading Windfalls for Wipeouts,* PLANNING MAGAZINE, (Chicago: American Society of Planning Officials, September 1974), pp. 9–13. *See also* DONALD G. HAGMAN AND DEAN MISCZYNSKI, EDS., WINDFALLS FOR WIPEOUTS: LAND VALUE CAPTURE AND COMPENSATION (Chicago: American Society of Planning Officials, 1978), esp. chap. 23.

15. 46 U.S.L.W. at 4865, 8 ENVIR. L. REP. at 20536.

- Where there is no present shading, but shading would occur if structures were built up to present zoning limits. For example, an area might be zoned for five stories but only built to two; two stories might be compatible with solar access requirements, but five might not, at least for good access on the existing two-story buildings.
- Where the authorized but unused density is in a zone adjoining the zone with solar units but close enough to block solar access. (These situations are discussed in detail in Chapter 1.)

A police-power-based TDR scheme would presumably be one in which a government downzoned an area or otherwise restricted the development rights of some lots to protect solar access and enabled affected landowners to transfer unused development rights to other appropriately located lots. Assuming the basic constitutionality of the TDR mechanism,[16] would its use for access protection conflict with the public purpose test? The answer is clearly no: courts give deference to legislative judgments respecting public health, safety, and welfare, and legislative and popular recognition of the need for conserving the nation's energy resources would seem an ample foundation for this innovation.

Although the 1954 U.S. Supreme Court decision *(Berman v. Parker)* virtually identified public use with public purpose,[17] a TDR solar access program based on eminent domain (rather than the police power) might not satisfy the public use test. For example, to ensure that solar access would be preserved in cases where a downzoned landowner refuses to voluntarily participate in the transfer scheme, a government might condemn his unused rights and put them in a public development rights bank for eventual resale to recoup its condemnation expenses.[18]

TDR landmark preservation programs along these lines arguably satisfy the public use criterion.[19] The applicability of the arguments to TDR solar access programs, however, is hindered by certain differences. In TDR landmark schemes, the restricted owner (landmark owner) receives either the TDR option (under a police-

16. *See* Costonis, *Chicago Plan,* passim.

17. *See* Chapter 3.

18. *See* Costonis, Space Adrift, chap. 2.

19. *See* Costonis, *Chicago Plan,* pp. 603–11.

power-based program) or actual cash from the city (if under a condemnation banking alternative) and a developer elsewhere ultimately funds the transaction by buying the transferred rights either from the landmark owner or the bank. In the solar access case, on the other hand, there is another participant: the benefited owner of solar collectors.

Two considerations favor the public use constitutionality of landmark TDR programs. First, the restricted owners (landmark owners) should, and in some cases must, receive benefits in dollars or in kind to protect the programs from the charge that they exceed the police power's scope by drastically reducing the value of the restricted owners' lands.[20] Second, a plausible argument can be made that it is fair to skew zoning levels at the transferee sites (by authorizing lower as-of-right[21] densities) because the threat to the landmarks' continued existence comes from activity in private land markets.[22]

Both propositions must be qualified for a TDR solar access program. The downzoned landowner may be similar to the landmark owner, in that the former may also suffer a loss of constitutional dimension so the government's use of its condemnation power in a scheme ultimately intended to compensate that loss serves a public use. But how can the boon given to the collector owners who benefited be justified if the city is in effect acquiring an access easement for them through its use of the eminent domain power? As indicated in Chapter 3, some courts would not permit such actions where the benefits of eminent domain are enjoyed by individual, identifiable members of the community. Other courts would permit them where they serve a public purpose; the trend appears to be to expand "public use" to be synonymous with "public purpose."

Planning issues. Without careful planning and certain economic preconditions, TDR will not work. Decisions on the shifting of densities to protect solar access must be consistent with other community goals and respect the requirements of sound planning. TDR should not be allowed to function as the tail that wags the planning dog. At least two key assumptions must be made if TDR is to be considered useful for solar access protection: the first is that it makes sense in general planning terms to downzone entire areas to secure

20. *See* Costonis, Space Adrift, chap. 1.

21. "As-of-right" means those development rights already available to developers under current zoning laws.

22. *See* John J. Costonis, *Development Rights Transfer: An Exploratory Essay,* Yale Law Journal 83 (1975): 75, 100.

access protection. While this may hold for low density residential neighborhoods, it may not for a central business district or an area of mixed uses undergoing intensive redevelopment that promises to contribute significantly to the community's economic health.

The second assumption is that it must also make planning sense to permit the transfer of densities to the transferee district. If the number and types of transfers are known, a reasonable assessment can be made of the program's public and private costs, its planning–administrative feasibility, and its potential legal impediments. But this could be impossible if transfer is indiscriminately opened to an indeterminate number of lots within the jurisdiction. For that reason, it would be unwise to use TDR for solar access protection unless the magnitude of anticipated transfers as well as the identity and relative percentages of landowners benefited and burdened can be estimated. Furthermore, the solar access protection implemented should be consistent with the planning characteristics and values of the affected community and its administrative capabilities.

If all sites within the transferee district were made eligible for receipt of the transfered density, the transfer of density may not solve the overall problem of protecting solar access but merely move it from the sites adjacent to the transferor site to the sites adjacent to the transferee site. TDR could be most useful if transferee districts can be found that, in addition to meeting the other planning and economic preconditions, are also areas where solar collectors are not expected. Two other potential transferee districts would be areas where existing densities were reduced (downzoned) so that TDRs would have to be purchased to attain what used to be as-of-right densities; or low density areas where both private market and public planning considerations justify a substantial increase in allowed densities (upzoning). However, under the first alternative, landowners could be expected to protest what they might view as an artificial manipulation of existing development rights to create a market for the transferred rights. (On the other hand, political protest could also be expected if the additional TDR densities were simply dumped into such areas without a prior downzoning.)

Economic issues. Two key questions concerning TDR's economic feasibility must be answered. The first is how much compensation, whether in kind (TDRs) or in dollars, should be given the landowner whose property has been downzoned to protect his neighbor's solar access: what is his actual loss as measured by the difference in the value of his property before and after downzoning and what proportion of this loss should or must the community restore?

Some TDR programs take the simple expedient of authorizing restricted landowners to transfer all their previously authorized but unused development rights.[23] This approach, which may be easier to administer, has problems because land values in the transferee district may be more or less than those at the transferor site.[24] Permitting transfers on a one-to-one basis from a low- to a high-value zone gives the restricted owner a windfall; in the reverse case, he suffers a wipeout.

In addition, it is not clear that government must give compensation equivalent to the full loss suffered under the restriction. Since landowners are constitutionally entitled only to some reasonably beneficial use return, the measure could be the difference between the return possible under a reasonably beneficial use and under the solar access standards. This approach would be more complex than the authorized-but-unused formula. The complexity may cause many jurisdictions to trade off the advantage of adjusted compensation against the administrative ease of the authorized-but-unused formula, especially if a large number of properties will be restricted.

The second key question is whether transferee districts will provide an active market for the transferred development rights. This turns on the demand for new construction of the type permitted under the TDR authorization and the currently allowed densities in the transferee districts.[25] However, there is no apparent relationship between solar use and the general health of a community's construction market, as there typically will be in the case of landmarks preservation. In addition, relatively few landmark owners would be expected to transfer rights under a landmark preservation plan while a potentially vast number of landowners (any owners whose construction would interfere with protected solar access) might do so in the TDR solar access alternative. Increasing the number of potential sellers increases the risk of saturating the market with TDR increment densities, thereby deflating the rights' value or requiring drastically low as-of-right densities in the transferee district to ensure a market.

Another problem is that if developers can acquire development rights free or for lesser payments, they will not buy rights from a

23. The New York City TDR-landmarks program is illustrative. *See* Costonis, *Chicago Plan,* pp. 603–11.

24. For an inventory of the various factors influencing the value of downtown parcels, *see* Costonis, Space Adrift, chap. 3.

25. *See* ibid., chap. 4.

public TDR bank or from transferor owners. Among the non-TDR sources of additional rights are variances, special exceptions, upzoning, or zoning bonuses (additional allocation of rights as a *quid pro quo* for the developer's provision of such amenities as plazas, arcades, and the like).[26] Hence, before a community institutes any type of TDR program it must first assess its policies and substantive decisions on these other mechanisms. If necessary, it must then modify these policies and decisions by limiting the amount of density available under them to levels that ensure a market for TDR-increment densities.[27] This is a difficult job both in its planning and political dimensions, but it is essential to the success of any TDR mechanism.

Conclusion

ELI lacks the data on which to base a firm conclusion on the utility of TDR as a tool for solar access protection. There is a need for precise planning–economic models; even in non-solar-access contexts, TDR is still somewhat experimental. State enabling legislation for solar access protection should, however, authorize local experiments with TDR as there may be situations where it would be useful.

Zoning by Special Assessment Financed Eminent Domain

Zoning by special assessment financed eminent domain (ZSAFED) is such a rare land use control that there is hardly any literature on it,[28] and it has been used in only three states (Missouri, Minnesota, and Wisconsin).[29] The essence of this device as it has been used in the past is that a majority of landowners in an area are given the power to create an exclusive residential district by using eminent domain powers to acquire rights to develop in any other

26. Costonis, *Chicago Plan,* pp. 573–79.

27. An analysis of the relation between the city of Chicago's bonus zoning provisions and the TDR provisions of the Chicago Plan and suggestions for how cities might accommodate TDR programs with other density-allocating measures of their zoning codes can be found in COSTONIS, SPACE ADRIFT, chap. 4.

28. It is not discussed, for example, in the standard treatises on land use planning law, such as ROBERT M. ANDERSON, AMERICAN LAW OF ZONING, 2d ed., 5 vols, (Rochester, N.Y.: Lawyers Co-operative Publishing Co., 1977), or NORMAN WILLIAMS, JR., AMERICAN LAND PLANNING LAW, 5 vols. (Chicago: Calaghan & Co., 1975). Consequently, the following remarks are based largely on Donald G. Hagman, *Zoning by Special Assessment Financed Eminent Domain (ZSAFED),* in HAGMAN AND MISCZYNSKI, WINDFALLS FOR WIPEOUTS, chap. 22, pp. 517–31.

29. *See* Hagman, *ZSAFED,* pp. 518–24 for ZSAFED's history.

manner. Instead of the government paying for the condemnation, all landowners in the ZSAFED district are assessed.

When ZSAFED was first used in 1893 in Kansas City,[30] controls banning nonresidential uses or particular residential uses (apartments, for example) from a district were not yet recognized as within the police power. Compensation for loss of development rights was necessary in order to do this, and assessment of costs against those benefited was reasonable.[31] Such controls are no longer a problem under the police power; zoning normally defines lawful uses for each district and often creates exclusive residential areas.

For solar access protection, a state could authorize a majority of landowners in an area, by vote or by petition, to form solar access ZSAFED districts. (To ensure coordination with local land use plans, the consent of the local planning authority should also be required.) This action would allow a majority in the district to gain the equivalent of a restrictive covenant for solar access and to compel those that would otherwise be holdouts from voluntary restrictive covenants to be bound by the will of the majority.

A peculiarity of this scheme in the past has been that the assessments would not be just against those who stand to benefit or who voted or signed the petition to form the ZSAFED district. They would also be levied against those who stand neither to be benefited or harmed, those who did not favor the ZSAFED, and even those who have been harmed and would receive compensation. Persons suffering losses would be paid their losses less their share of the assessment for compensation. An alternative scheme could be to assess only those who benefit from the restrictions (those who had solar collectors or good collector sites) or only those who favored forming the solar access ZSAFED district.

The amount of compensation has been figured as it is in eminent domain cases—the full drop in value of the property due to the restraints imposed. An alternative, when the restraints could be imposed in whole or in part under the police power, would be to limit the compensation to some substantial fraction (such as one-half) of the loss exceeding some threshold amount (such as one-tenth of the land's former value).[32]

30. Ibid., p. 518.

31. For example, ZSAFED was first used in Minnesota in 1915 after an earlier attempt to zone commercial buildings and apartments over two and one-half stories out of a residential area of Minneapolis was struck down as not a proper exercise of the police power. Ibid., p. 519.

32. Thus, the compensation for a drop in value from $50,000 to $40,000 might be

Assessments in the past have had to cover the fees of appraisers specially appointed to evaluate the losses in a particular ZSAFED district. Presumably, the appraisals could be done less expensively by the government's property tax assessors, who would have to reappraise the property for tax purposes anyway. Appraisal costs could thus be underwritten by a government wishing to encourage solar energy use.

Other variations on this basic ZSAFED scheme in a nonsolar context have been suggested,[33] but do not merit discussion here. ZSAFED, like TDR, appears to be a tool for solar access protection that could be useful under some circumstances, particularly where a community wishes to impose growth restraints on development that could not be fully justified under the police power or that would not be in accord with the local sense of fairness or political acceptability if there were no compensation. In such cases, if a majority of landowners in an area were willing to form a ZSAFED district, necessary restraints could be imposed and compensation made by affected landowners, rather than by all taxpayers in a locality.

For this reason, ELI recommends that state enabling legislation authorize communities to permit experiments with the ZSAFED technique for solar access protection.

$2,500, or half of the difference between the loss ($10,000) and one-tenth of the original value ($5,000).

33. Hagman, *ZSAFED,* pp. 524–30.

Part V
Implementation

Previous chapters have dealt with the "what," "how," and "when" of solar laws. These chapters concentrate on the "who." Who will make the policy decisions that form the law—state, local, or federal officials? Who will implement various parts of the law—zoning boards of appeals, building departments, planners or others? Who will make enforcement decisions—administrators, mediators, arbitrators, or judges and juries?

These issues are not unique to solar access. They arise when any new law is written. But because solar access protection represents a new field of law that may be unusually sensitive to public opinion, the officials involved can have a major impact on reactions to these laws. For example, dispute resolution methods are particularly important since they will touch the public directly. Similarly, the source of solar access laws—whether they spring from well-defined local or regional needs, or are imposed from the outside—will also influence public opinion.

Most laws affect most people only indirectly: Antitrust laws, security regulations, or patent laws, for instance, are more of a backdrop to daily living than part of the script. Solar access ordinances belong to another, smaller group of laws—such as zoning, traffic, and tax rules—that directly affect almost everyone.

12 Local, State, and Federal Roles

Solar access protection is a concern at more than one level of government. The following decisions could be made at the federal, state, or local level:

- □ What legal device should be used for solar access protection?
- □ How much protection should be given?
- □ Should protection vary with individual sites or area-wide characteristics, and if so, how?
- □ Who should approve individual applications, either for solar access permits or development permits affecting solar access?
- □ What kinds of remedies should be available?
- □ How much discretion should administrators have?
- □ What lands should receive solar access protection?

Local Roles

Land use controls have traditionally been exercised by local subdivisions of states, and it is in local governments that the mechanisms for administration and enforcement of land use planning and controls are most developed. This is a strong argument for giving localities the primary responsibility for administering and enforcing solar access protection laws. There are other reasons for preferring local responsibility: cities and counties are the most familiar with particular facts relating to land use patterns and needs in their communities, are the most directly accountable to the persons who will be affected by access controls, and are the most likely to be able to strike the optimum balance between protecting solar access and allowing necessary development. (Some localities do not regulate land use, but this usually occurs in rural areas where solar access protection laws will seldom be needed.)

Finally, there are many local factors to be considered when an access law is drafted or contemplated: topography; height, bulk, and location of structures and vegetation; present and proposed land use controls and comprehensive plans; latitude; climate; energy use of structures; and orientation of streets, roofs, and walls. The consideration of such local particulars is essential to ensure that a law's protection is neither too narrow nor too broad.

On the other hand, several arguments can be made against leaving decisions on solar access entirely to counties or municipalities. The most compelling of these seems to be that many localities lack the necessary expertise and funding to implement solar access protection. ELI recommends that federal and state governments provide that funding and expertise when it is lacking. It would be particularly helpful if centralized governments provide the basic research and development for use in all localities. For example, they could develop general programs for calculating available sunlight and provide a training seminar for local officials all over the country.

In addition, some localities may be dependent on the tax revenues that development attracts and may be biased in favor of intensive development. Developers are often active in local politics and may work against solar access laws they believe would restrict their livelihoods. On the other hand, a locality may be so antagonistic to development that it uses solar access protection as a pretext for exclusionary zoning. Other local governments may adopt solar access ordinances that do not regulate their own, municipal, activities (such as public tree maintenance or municipal building construction). While any of these events could occur, the mere possibility does not justify federal or state intervention. Existing remedial mechanisms (such as court decisions and the election of new local officials) should be tried first.

ELI believes that localities are best suited to answer the questions posed at the beginning of this chapter.[1] The issue is whether

1. An economist concluded several years ago that local flexibility should be favored

they would do so. Once the states enact needed enabling legislation and the federal and state governments develop adequate financial and technical support mechanisms, it will quickly be apparent whether local solar access ordinances are being adopted, implemented, and enforced as they should be.[2] If localities fail to adopt needed solar access legislation, a more active state or federal role could then be taken.

State Roles

Preempting

Ultimately, the police power resides in the states, and localities have a power to legislate for solar access only to the extent that the states have—by constitution or legislation—delegated power to the localities. Whether or not they delegate this power, states could themselves adopt solar access legislation and preempt local solar access rules. This would not necessarily involve state implementation and enforcement; local governments could be instructed to carry out those responsibilities under the state law.

The Massachusetts bill discussed earlier[3] is of this kind, preemptive, but using local enforcement. Although the Massachusetts bill does not by its terms preempt different, more comprehensive solar access protection by localities, it is so pervasive a scheme that it could be construed as preemptive by implication. Such an approach may be feasible in a state like Massachusetts, which is largely made up of incorporated towns and where development patterns are said to be fairly uniform.[4] However, many states have large unincorporated areas for which the state would have to provide the enforcement machinery if statewide enforcement were to become a reality. (In other states, county governments are given strong powers in sparsely settled areas.) Local conditions often differ widely, even in small states such as Maryland. If a solar access law is to be fair and effective, it must be adapted to local conditions. Thus, either the statewide legislation itself or the regulations adopted thereunder

over national uniformity in designing environmental laws if local governments were free of undue political pressure. *See generally* Richard O. Zerbe, *Optimal Environmental Jurisdictions,* Ecology Law Quarterly 4 (1974): 193–245.

2. *See* discussion of Oregon experience, Chapter 3, notes 20 and 21 and associated text.

3. *See* Chapter 7, at notes 7 and 8 and associated text.

4. From a conversation with William C. Osborn, former director of the Massachusetts Solar Action Office (since abolished), which drafted the Massachusetts bills; whether there is great enough uniformity for satisfactory statewide solar access protection remains to be seen.

must be fairly complex or unnecessary hardships will be imposed on the affected residents. A preemptive state solar access law might be effective even where statewide conditions are not uniform if it only applied to the more developed areas (exempting rural areas, for example); if it had different standards for different developmental, geographical, and political conditions; and if it delegated enforcement to localities. A related approach is to allow local jurisdictions to "opt out" of a state plan.[5]

Mandating

Instead of preempting local solar access ordinances, a state could require localities to adopt such ordinances and could specify standards the ordinances must meet; it might also provide for state review of proposed ordinances and regulations.[6] There may be some states that are compact, densely developed, and homogeneous enough that mandating statewide solar access protection would make sense. In addition, the state could monitor enforcement and provide for state reversal of unsatisfactory local rules and appeals. Presumably, different standards could apply to different kinds of localities. Rural areas could be exempted. (In the few cases where access protection is needed in these areas, private agreements could be attempted.) Where a locality failed to adopt and enforce solar access protection satisfactorily, the state could either provide more detailed guidance to the locality or act on the locality's behalf. State review could be limited to approving or disapproving local rules or could include substituting new rules. (Precedents for this monitoring form of legislation abound on both the federal and state levels—examples are state sanitary codes and the federal Coastal Zone Management Act.)[7] Or a state could choose the legal device or devices to be used and set minimum and maximum standards to limit local discretion while allowing localities to adapt the device to local conditions.

5. *See,* e.g., 1978 Cal. Stats. ch. 1366, § 1 (to be codified as CAL. PUB. RES. CODE § 25985) (passed 31 August 1978).

6. Daniel R. Mandelker notes that state-level reviews of local decisions may be frustrated by the inadequacies of the decision-making process at the local level. He suggests three forms that state review may take: (1) a state agency may function as an appellate body; (2) review may be independent of an appeals process (allowing a commission to modify inconsistent plans or call in important local land development decisions on its own initiative); (3) a state agency may be authorized to review the content and enforcement of controls as a condition of state assistance. ENVIRONMENTAL AND LAND CONTROLS LEGISLATION (Indianapolis: Bobbs-Merrill, 1976), pp. 26, 20.

7. 16 U.S.C. §§ 1451–1464 (1976).

Enabling

The need for enabling acts to broaden local powers so that cities and counties can regulate both vegetation and structures specifically for solar access is discussed in Chapters 1, 4, and 7. Enabling legislation need not leave all decisions to local discretion. It may limit both the devices authorized and local discretion, just as mandating legislation would. Enabling legislation that shows a strong public purpose for access legislation, together with the specific authorization of such devices as solar collector recordations, envelope zoning, solar overlay zones, and vegetation controls, would go far toward protecting local solar access ordinances from attacks based on an alleged lack of public purpose, improper means of implementation, or improper subject matter for zoning.[8]

States can also play an important role in providing remedies to support solar access legislation. They can provide for injunctive remedies in state courts of general jurisdiction. Such remedies are essential if solar access is to be preserved, rather than traded for damages. Specific authorization from the state would also clarify a locality's authority to use other remedial provisions such as requiring that vegetation be trimmed, allowing the city to enter and trim vegetation, shifting the burden of proof, or specifying damages.

One of the principal criticisms of zoning as a solar access protection device is that in many jurisdictions zoning can easily be changed. Because this could leave collector owners suddenly without protection, the likely effect would be to discourage potential solar energy users. However, state zoning enabling acts can limit the conditions under which zoning changes can be made. For example, the Standard State Zoning Enabling Act provides that:

> Such regulations, restrictions, and boundaries may from time to time be amended, supplemented, changed, modified, or repealed. In case, however, of a protest against such change, signed by the owners of 20 percent or more either of the areas of the lots included in such proposed change, or of those immediately adjacent . . . , such amendment shall not become effective except by the favorable vote of three-fourths of all the members of the legislative body of such municipality.[9]

8. It would also be useful, as with mandating legislation, to include language in the enabling legislation that prohibits local governments from using these same powers to unreasonably restrict the installation of solar equipment.

9. U.S. Department of Commerce, Advisory Committee on City Planning and Zoning, A Standard State Zoning Enabling Act (Washington, D.C.: U.S. Government Printing Office, 1926), § 5. The act is reproduced in full in Robert M. Anderson, American Law of Zoning, 2d ed., 5 vols. (Rochester, N.Y.: Lawyers Co-operative Publishing Co., 1976), 4: § 30.01, pp. 398–404.

A similar provision could require this type of procedural safeguard whenever a certain number or percentage of solar collector owners or persons in a protected area object to a proposed decrease in the solar access protection. The specific percentages and numbers could be adjusted to reflect local requirements.

Supporting

If localities are to have a significant role in solar access protection, they may need considerable financial and technical assistance from regional planning agencies, states, or the federal government.[10] States should direct existing state energy and land use planning agencies to acquire the necessary expertise to provide local governments with technical assistance and to administer grants for necessary local planning and implementation of solar access ordinances.

While ELI believes it would be appropriate for the federal government to develop and fund the needed technical services, the states could deliver those services more effectively. The laws of the states vary just as local topography and climates vary. The relationships between states and their subdivisions can affect the kinds of options realistically available for solar access protection. (For example, a county that has no zoning power cannot use zoning-based controls.)

States should consider where solar access controls are needed and feasible and whether the authority to exercise them is available. Of the roughly 60,000 jurisdictions with zoning power, only about 5,000 of them exercise it;[11] and of the jurisdictions with power to adopt building codes, only about 8,000 of them have.[12] It should not

10. Reviewer Donald G. Hagman dissents from this opinion and notes that state and federal assistance may make solar access protection appear less expensive than it actually is. He also questions the efficiency of this approach. Reviewer James Matthews similarly commented that the real expertise is at the local level and that the costs should be borne by those who are benefited. ELI agrees that local financing is best, where possible, but believes that the benefits of using solar energy will be shared by the whole nation, making federal or state support acceptable. Further, while localities may have superior expertise on land use controls as applied to their communities, some relevant information on the effect of solar access and solar energy systems on land use can be gathered more cheaply and efficiently in a centralized manner. A letter (dated 11 October 1978) from Mark Plaat, energy planner with the Albany County, N.Y., Planning Board, addressed to "all interested parties," states that "Planning Boards lack the technical expertise to properly evaluate the land use impacts which result from alternate energy development." He calls for state and federal help.

11. As of 1974. PETER WOLF, THE FUTURE OF THE CITY: NEW DIRECTIONS IN URBAN PLANNING (New York: Watson-Guptill Publications, Whitney Library of Design, 1974), p. 149.

12. As of 1975. CHARLES G. FIELD AND STEVEN R. RIVKIN, THE BUILDING CODE BURDEN (Lexington, Mass.: D.C. Heath and Co., Lexington Books, 1975), p. 33.

be assumed, however, that the areas where zoning and building codes have not been adopted should do so just to provide for solar access protection. Most of these places are rural or exurban areas with sparse development and relatively low energy consumption where privately acquired solar easements should suffice.

States should determine what technical support and legal advice will help localities make the best use of existing agencies and legal institutions. Although funding for such efforts could be provided by the federal government, the cost may be within the means of some state governments. A bill to provide federal support for state solar access studies is discussed later.[13]

States should enact legislation that makes it clear that easements for solar access are a valid interest in land that may be created, transferred, and recorded like any other interest in land (see Chapter 10).

States can give localities what, in effect, would be the power to regulate land use for purposes of solar access protection over the state's own property. Ordinarily, state property is immune from local control,[14] just as federal property is immune from state and local control.[15] State land use in urban and suburban areas could cause serious problems for solar access unless coordinated with local (or state) solar access legislation. For this reason, both prototype ordinances in this report provide that a state cannot enjoy solar access protection unless it consents to be bound by the solar access ordinance's rules. States need not literally delegate control over state property to localities, but they can require that all state construction and vegetation comply with local ordinances designed to protect solar access. (Similar federal legislation may be advisable.) To avoid being bound by unreasonable local ordinances, a state may wish to set criteria the ordinances must meet if they are to bind the state. These criteria could be part of the same legislative package that enables or requires localities to adopt solar access ordinances. In any event, states should consider requiring local governments to abide by any solar access controls they enact.

Conclusions

Except where conditions are uniform statewide, ELI considers state preemptive legislation ill-advised. The choice between mandating and enabling legislation depends on states' evaluation of the

13. S. 3027, 95th Cong., 2d Sess. (1978). See notes 46 and 47, below, and accompanying text.

14. Robert M. Anderson, American Law of Zoning, 2: § 12.06, p. 408.

15. Ibid., § 12.07, p. 416.

probability that localities would enact and enforce satisfactory solar access ordinances, where needed, on their own. In general, it would seem best for states to play an enabling role.[16] Localities could then experiment with various approaches. The need for solar access protection can be expected to grow rapidly. Thus if localities do not act, once enabled to, states could require them to do so.

Although all of the following actions will not be necessary in every jurisdiction, states should nonetheless enable localities to:

- □ Adopt or revise local comprehensive plans to encourage or require solar energy use (this action should probably be required)
- □ Enact specific regulations protecting solar access in existing developed areas as well as new developments
- □ Use zoning powers to provide for solar access and use the police power generally to provide for solar collector recordations
- □ Enact specific subdivision regulations requiring private developers to provide for solar access in both layout of developments and restrictive covenants
- □ Obtain (and record on the land records) solar access easements by purchase, gift, or condemnation for either private or public solar installations[17]
- □ Use transferable development rights and zoning by special assessment financed eminent domain to protect solar access (TDRs and ZSAFED will probably not often be used but should be allowed)
- □ Control vegetation through such regulations as zoning, subdivision, and public nuisance controls
- □ Provide for injunctions, damages, trimming, arbitration, and other remedies as needed to enforce solar access controls.

16. *See generally* ERIC STRAUSS AND JON KUSLER, STATUTORY LAND USE CONTROL ENABLING AUTHORITY IN THE FIFTY STATES (U.S. Department of Housing and Urban Development, Federal Insurance Administration, HUD-FIA-179, September 1976).

17. SEE A FORUM ON SOLAR ACCESS: PROCEEDINGS OF A FORUM ON SOLAR ACCESS HELD BY THE NEW YORK STATE LEGISLATIVE COMMISSION ON ENERGY SYSTEMS (Rockville, Md.: National Solar Heating and Cooling Information Center, 1977), app. 2.

A state may also wish to provide for a presumption of validity when areas are downzoned for solar access protection purposes, even in areas where solar energy installations are not anticipated at this time. (In some states, such as Maryland, there is a presumption against such rezoning.) If downzoning for solar access were to become needed, a state would be prepared.

Federal Roles

Extent of Federal Powers

The federal government has only those powers provided to it in the U.S. Constitution; all other powers of government are specifically reserved for the states.[18] Except for federal lands, the power to regulate the use of land directly is not explicitly given to the federal government.[19] Nonetheless, a plausible case can be made that the U.S. government has the authority to protect solar access on all land in the nation through its powers to regulate interstate and foreign commerce, collect and spend funds, provide for the common defense, provide for the general welfare, and make all laws necessary and proper for executing the enumerated powers of the federal government.[20] Although the following discussion argues that these powers exist, ELI does not recommend that they be used for solar access protection except in the event of a severe national energy emergency.

The power to regulate interstate commerce is perhaps the broadest of the federal powers that could be brought to bear on land use.[21] It is generally the case that any conduct that may, in itself, be trivial but that collectively has a substantial effect on interstate commerce can be reached by federal regulations under the commerce clause.[22] While the contribution of individual solar energy installations to decreased use of other energy sources in (or affecting) interstate

18. U.S. CONST. amend. X.

19. Ibid., art. IV, § 3.

20. Ibid., art. I, § 8.

21. *See generally,* Comment, *State Environmental Protection Legislation and the Commerce Clause,* HARVARD LAW REVIEW 87 (1974): 1762–85.

22. In a classic case, Wickard v. Filburn, 317 U.S. 111 (1942), the federal government was able to prohibit a farmer from growing more wheat than was his allotment, even though the wheat was grown on his own land and fed to his own livestock. It did not matter that there was no interstate commerce directly involved in his action, nor that his individual contribution to the demand for wheat he would otherwise have to buy was trivial, since his contribution—taken together with that of many others similarly situated—was far from trivial.

commerce would be trivial, the cumulative effect of many such installations could be substantial. Although the conduct to be reached by solar access legislation is that on adjacent lands rather than that of the solar energy system user, it should be possible to reach that conduct too.

> Congress . . . has unquestioned control over any business activity which in any way "affects" commerce, regardless of how "local" its effect may be, and how small or insignificant the contribution of a single instance may be if it is representative of "many others similarly situated."[23]

The building industry uses many materials in interstate commerce and has a significant impact on the national economy. Reaching construction that would impede solar access should be possible under the commerce clause. Furthermore, solar devices would compete with other energy sources such as oil and gas that move in interstate commerce and would therefore be subject to congressional regulation. Reaching noncommercial vegetation, especially that on properties of individual homeowners, may not be possible.

Other enumerated federal powers could also extend to solar access protection. For instance, the federal government is responsible for regulating foreign commerce. Reducing imports of gas and oil would improve our balance of payments situation. It would also improve our defense capabilities. Reduced fossil fuel use would improve our air and water quality, which would be in the general welfare and would also reduce interstate movement of air and water pollutants. Roughly one-third of U.S. land and all of the outer continental shelf belongs to the federal government. Reduced use of conventional energy would help conserve the energy resources in these federal properties. All these effects could be expected from widespread use of solar energy, but widespread use of solar energy may not occur without adequate solar access protection. Federal action to protect solar access could thus be seen as a necessary and proper use of the federal government's enumerated powers.[24]

Preempting

Any preemptive bill could be recast into a mandatory form (see below), which would be preferable but still not recommended.

23. Paul R. Benson, Jr., The Supreme Court and the Commerce Clause, 1937–1970 (Cambridge, Mass.: Cunellen, 1970), 345–46.

24. *See* Philip Soper, *The Constitutional Framework of Environmental Law,* in Federal Environmental Law, Erica L. Dolgin and Thomas G. P. Guilbert, eds., (St. Paul: West Publishing Co., 1974), p. 20, for further discussion of related constitutional issues.

Assuming that federally proposed levels of protection would be generally effective, it would be better to mandate state or local solar access at some minimum level of protection and leave the option open for more stringent controls that would provide better access where local conditions permitted. A preemptive approach would foreclose opportunities to do more than what the federal government provides.[25]

Federal legislation might provide for national solar access protection with administration and enforcement by federal agencies and courts. This has not, apparently, been seriously proposed. The federal government has in the past generally relied on states and localities to administer and enforce environmental legislation relating to land use. An attempt to centralize enforcement of land use controls on a parcel-by-parcel basis through the federal government would almost certainly be politically doomed.

Federal preemption is desirable only when there is a need for national uniformity. While some minimum solar access may be needed over much or all of the country, uniformity is less important than adaptability to local conditions. There is even greater diversity of developmental, topographic, climatic, and other conditions nationwide than there is statewide. It would be impossible to write a national preemptive statute or set of regulations suitable for all.

Mandating

The federal role in land use controls is growing.[26] In the early 1920s, the Department of Commerce promulgated the widely adopted Standard State Zoning Enabling Act.[27] Since World War II, Congress has passed many housing, urban development, and environmental bills containing planning components.

Mandating federal legislation might require state or local solar access protection with federal review of the means adopted and with federal rules, administration, and enforcement only if the state or

25. For example, federal preemption in the nuclear power field has kept Minnesota from enacting more stringent controls on disposal of radioactive waste. Northern States Power Co. v. Minnesota, 447 F.2d 1143 (8th Cir. 1971), *aff'd*, 405 U.S. 1035 (1972).

26. Two recent and important examples with a mandating federal role are the Clean Air Act, 42 U.S.C.A. § 7401 *et seq.* (1978), and the Federal Water Pollution Control Act, 33 U.S.C.A § 1251 *et seq.* (1978).

27. U.S. DEPARTMENT OF COMMERCE, ADVISORY COMMITTEE ON CITY PLANNING AND ZONING, A STANDARD STATE ZONING ENABLING ACT.

locality failed to adopt acceptable means.[28] Such legislation could give considerable discretion to states or localities to adapt the control to their needs. The principal technical disadvantage of the preemptive approach—inflexibility—would thus be overcome. But political resistance to such legislation would be great because it invades what is traditionally a matter of local and state control and goes far beyond the impact of other environmental legislation.[29]

The suggestions that the federal government compel states to act in some manner raises fundamental questions concerning federalism. The states are perceived as being autonomous units within a unifying national government, and the Constitution restricts the manner in which the federal government uses state governments to further national goals.

A section of the Social Security Act of 1935 created assistance programs that would be administered by the states in accordance with federal standards. In *Steward Machine Co. v. Davis,*[30] the Supreme Court upheld this provision against a claim that it constituted a direct attack on the autonomy of the states. However, two recent Supreme Court cases have set this area into considerable turmoil.

The first case arose after the Environmental Protection Agency promulgated regulations under the Clean Air Act to control air pollution by automobiles. Because the federal government could not hope to enforce regulations as complex as these, the EPA required states to implement the national standards or to set their own control criteria. Congress did not directly authorize EPA to act in this manner. Not all states readily accepted this mandate, and four courts of appeals heard cases challenging the regulations. As expected, the conflict found its way to the Supreme Court where, in *EPA v. Brown,*[31] the Court remanded the cases because the federal government agreed to delete requirements that would require the states to adopt the regulations. Thus, the Court did not rule directly on the point of federally mandated action.

28. Federal legislation might also require the consideration of solar access impacts in federal environmental impact statements (see Chapter 7).

29. The federal government has mandated state action in a variety of statutes over the past several decades (see note 26). Another example is the Coastal Zone Management Act of 1972.

30. 301 U.S. 548 (1937).

31. 97 S. Ct. 1735 (1977).

In *National League of Cities v. Usery (NLC),*[32] the Supreme Court held that federal requirements that state and local governments adopt minimum wage and maximum hour standards for their own employees constituted an invalid invasion of state and local government autonomy under the Tenth Amendment. The holding in *NLC* pertained only to those "functions essential to separate and independent existence"[33] of the states and to the "integral portion of those governmental services which the States and their political subdivisions have traditionally afforded their citizens."[34] It is conceivable that if the federal government attempted to protect solar access by fundamentally revising concepts of property law, the attempt would run counter to this holding. (ELI does not believe it is necessary to make fundamental changes to property law in order to protect solar access.)

An extensive analysis of the environmental law implications of *NLC* led one commentator to conclude that it is "most unlikely that recognition of constitutional power in Congress to require state implementation of federal environmental requirements would pose a serious threat to state political autonomy"[35] but that "the values of local self-determination are surely as important in matters such as zoning and pollution control as they are in the delivery of protective services, education, sanitation, maintenance of parks and recreation facilities, and other functions deemed integral by *NLC*."[36]

Rather than directly mandate certain state actions, federal legislation might provide that state and local laws are invalid to the extent that they permit development that would adversely impact upon solar access. This has been the approach of the poorly conceived federal solar access bills introduced so far.[37]

This type of legislation is not recommended partly because it affords state and local governments no opportunities to exercise reasonable discretion and planning by balancing development against

32. 426 U.S. 833 (1976).

33. 426 U.S. at 845.

34. 426 U.S. at 855.

35. Stewart, *Pyramids of Sacrifice: Problems of Federalism in Mandating State Implementation of National Environmental Policy,* YALE LAW JOURNAL 86 (1977): 1264.

36. Ibid., p. 1267.

37. H.R. 1167, 94th Cong., 2d Sess. (1976); and H.R. 259 and S. 985, 95th Cong., 1st Sess. (1977). These are substantially identical bills.

solar access needs. It would nevertheless be reasonable for the federal government to closely monitor state and local actions to protect solar access. Should a national emergency arise requiring more extensive use of solar energy, a mandating role would become more reasonable, or even necessary if adequate access was not protected.

Supporting

In lieu of directly protecting solar access, the federal government could use its spending power to provide grants-in-aid or other incentives to states and localities to protect solar access or to provide technical assistance. The federal power exercised in this indirect way is more flexible and far-reaching than the power to regulate directly. There would be no question as to the validity of providing grants for developing and adopting state or local solar access laws regulating vegetation on private property. Legislation might provide grants for states and localities to adopt approved means of solar access protection.[38]

Threats to withdraw federal funding have had only limited success, however, and therefore have little credibility as a way to force state compliance with federal guidelines.[39] The practical consequences of withdrawing federal support from programs authorized by Congress would be self-defeating and the resulting voter resentment could not be overlooked. Provided there were no time limits in a federal statute, a state could proceed on its own to encourage use of solar energy and wait to apply for federal grants until a majority of its citizens felt that the benefits to be gained from the federal grants would be sufficient to overcome any perceived disadvantages associated with redefining property rights. Both prototype ordinances in this report, along with the kinds of adaptation to local conditions that are recommended, may require more money to develop than some localities will feel able or willing to spend.

As part of its supporting effort, the federal government could develop a program the states and localities could use to determine available solar access. It should be possible for such a program to

38. This would be analogous to the approach taken by the Coastal Zone Management Act of 1972, 16 U.S.C. § 1451 *et seq.* (1976), or to the federal assistance to state governments that develop energy conservation plans, 42 U.S.C. §§ 6322, 6323 (1976), and to the prohibition against federally insured or regulated financial institutions financing construction projects in states whose building codes do not meet federally mandated energy conservation standards, 42 U.S.C.A. § 6834 (1977).

39. Edward A. Tomlinson and Jerry L. Mashaw, *The Enforcement of Federal Standards in Grant-in-Aid Programs: Suggestions for Beneficiary Involvement,* University of Virginia Law Review 58 (1972): 619–29.

generate maps of solar access actually available (as opposed to that theoretically available when there are no trees and no other buildings near a subject building). Other maps could show the solar access that would be protected under alternative assumptions concerning types of controls. These alternative assumptions should include: (1) the existing land use restrictions and growth of vegetation without controls: (2) various possible levels of restrictions on structures and vegetation; and (3) restraints allowing essentially no new development and requiring some cutting back of existing vegetation (under eminent domain).

With such mapping in hand, a community could choose the restraints that are politically and economically acceptable. There would probably be cases where there is little to gain from severely restraining development and also cases where much access can be gained for little extra restraint. Mapping could also greatly strengthen arguments for a rational basis to controls and protect ordinances against legal attack for being arbitrary and capricious or for not offering equal protection.

ELI cannot estimate the cost of preparing the solar access components of comprehensive plans and mapping areas requiring different treatments but would recommend that states or the federal government examine the costs of alternative ways of accomplishing these tasks. Whatever the costs, localities are much more likely to undertake such activities when they have state or federal help on either a cost-sharing or fully funded basis. Funding could be tied to developing solar access protection schemes that meet some minimum standards. A precedent exists in the federal government's technical and financial assistance to state governments to develop energy conservation plans.[40]

HUD has a Section 701 Community Development Program, which finances state or regional land use planning that meets certain criteria.[41] Recipients must consider "projections of land use needs and land resource development including energy facilities siting needs."[42] This clearly includes central power plants, and arguably includes solar installations. Recipients must consider the "conservation of energy through land use strategies designed to reduce energy consumption and the development of policies designed to facilitate

40. 42 U.S.C.A. §§ 6322–6323 (1977).

41. Section 701 of the Housing Act of 1954, as amended, is codified at 40 U.S.C.A. § 461 (1978).

42. 24 C.F.R. § 600.72(2) (1977).

the recovery of energy resources in a manner compatible with environmental protection."[43] Protecting solar access could be viewed as an energy conservation program and as a policy to facilitate recovery of solar energy resources. Federal assistance to states and localities for development of solar access comprehensive plans and mapping (and other related work) could be funneled through existing programs,[44] through the regional solar centers, or through new programs in the U.S. Department of Energy.

A planned, comprehensive approach is preferable to the alternative, although it will have a much higher initial cost (much like the installation of solar equipment). The development of computer software to process necessary solar access data should be undertaken by the federal government. The basic method chosen should be universally applicable.

Much of the data on the amount of solar access that presently exists in a given locality could be gathered from aerial photographs, with machine processing to minimize cost and errors. The federal government should be able to develop the necessary procedures using existing techniques and equipment. It could then provide or fund the photographing and data processing, either directly or through a state agency, that would provide technical support for cities and counties.

Legal authority for technical assistance already exists. Federal law authorizes departments and agencies to provide special or technical services to state and local governments, but the state or local governments must generally pay for these services.[45] In certain cases, such as assistance to develop state energy conservation plans, no reimbursement is necessary.[46] Legislation along these lines should provide for free or subsidized technical services to states and localities to aid in developing comprehensive solar access plans and mapping, and to support similar state technical services to localities for those special areas where federal technical support is not sufficient.

To provide the kind of technical support recommended, the federal government will have to undertake some additional research. Although some work has been done on hourly solar energy available

43. Ibid., § 600.72(7) (1977).

44. HUD also has other financial and technical assistance programs, such as the Demonstration Cities and Metropolitan Development Program, 42 U.S.C.A. § 3301 *et seq.* (1977).

45. 42 U.S.C.A. § 4222 (1977).

46. Ibid., § 6323 (1977).

at a number of cities throughout the U.S.,[47] little information concerning whole-day radiation is available. Solar access legislation cannot be premised on protecting the whole day's sunlight, however, since this would prohibitively restrict development.[48]

Both for solar access purposes and for market penetration studies, a method for computing solar energy actually available at existing structures is urgently needed. ELI recommends that the federal government, with the help of the regional solar energy centers, begin to develop such a method as soon as possible.[49]

It is necessary to account for the effects of actual weather, topography, existing structures and vegetation (which would presumably be grandfathered into any controls), street and structure orientation, etc., to figure what fraction of sunlight is actually available. No reasonable estimate can be made of what the potential is for retrofitting existing structures with solar energy systems if these complicating factors are not considered. It is not known, for example, what the actual retrofit potential is for the nation as a whole, or even for any particular area.

In addition to funding and technical assistance, the federal government could encourage state and local solar access legislation by

47. T. Kusuda and K. Ishii, Hourly Solar Radiation Data for Vertical and Horizontal Surfaces on Average Days in the United States and Canada, National Bureau of Standards Buildings Science Series, no. 96 (Washington, D.C.: U.S. Government Printing Office, April 1977).

48. It is fairly easy to compute what fraction of the whole day's solar energy is available during a given period of a day, assuming clear atmosphere. One solar study has graphs summarizing the percentages of daily radiation available at a particular latitude in the months of December, March, and June for collectors at various tilts from vertical to horizontal as well as in several modes of tracking. *See* Melvin M. Eisenstadt, Stephen C. M. Long, and Albert E. Utton, *A Proposed Solar Zoning Ordinance,* mimeographed, prepared for the New Mexico Energy Institute (ERB-76-103, November 1977), Figures A-2 through A-16; these apply at 35° north. However, to know the maximum amount of solar energy that would be provided, say, by protection from 9 A.M. to 3 P.M. in any particular locality, it is necessary to know not just how much sunlight is available over a whole day at a particular time of year, but how much is available between 9 A.M. and 3 P.M. during the annual expected collector use period, taking actual weather into account. *See* Ibid. Kusuda and Ishii's work goes this far. But data are in tables, and fractions would have to be computed. Graphs like Eisenstadt and Utton's could be helpful. Another form of graphic presentation that could be helpful can be found in Note, Solar Energy 9, no. 3 (1965): 164.

49. Research should not be limited to finding the energy available during specified hours of the day but should also include energy available for a solar altitude to various parts of structures with specific height-to-spacing ratios. This last category could be based on typical situations found from examination of regional samples of developed areas.

binding or pledging itself to comply with such legislation. The federal government has the exclusive power to control its own land, including privately owned but government-leased land.[50] This power extends to federal property that is leased and used by private parties, unless the federal government has ceded control to state legislative jurisdiction. The Advisory Commission on Intergovernmental Relations (ACIR) has recommended since 1961 that Congress authorize and direct federal agencies to cede legislative jurisdiction to the states as rapidly and extensively as would be consistent with their program needs. While this more general action is not required for solar access protection, federal legislation is needed requiring all federal agencies to construct and maintain their facilities (including vegetation) in a manner consistent with state or local solar access law.[51]

Conclusions

The best approach to solar access protection is to let states and localities take the lead, but to bolster their efforts with federal financial and technical support. This would be in keeping with our federal system, under which the national government steps in only after states and localities have had the first opportunity to act but could not or would not. This also would advance the idea expressed by Justice Brandeis that:

> It is one of the happy incidents of the federal system that a single courageous State may, if its citizens choose, serve as a laboratory; and try novel social and economic experiments without risk to the rest of the country.[52]

In 1978, a solar energy assistance bill was introduced in the U.S.

50. U.S. CONST. art. IV, § 3. *See also* ANDERSON, AMERICAN LAW OF ZONING, 2: § 12.07, pp. 416–17.

51. Precedent exists for such legislation. The Lanham Act says that defense housing should be constructed after consultation with local officials so that projects can comply with local plans. 54 Stat. 1127, as amended (formerly codified at 42 U.S.C. § 1545.). The Federal Urban Land-Use Act requires the General Services Administration, when acquiring, using, and disposing of federal land, to be consistent to the "greatest extent practicable" with zoning and land use practices and planning and development objectives of the local government and planning agencies. 40 U.S.C. § 531 (1976). Similarly, the Coastal Zone Management Act (CZMA) has required federal agencies conducting or supporting activities directly affecting the coastal zone, or undertaking development in it, to do so in a manner that is to the "maximum extent practicable" consistent with approved state management programs. 16 U.S.C. § 1456(c) (1976). Under the CZMA, the federal government is in the position to approve the state plans before being obliged to follow them; it may be similarly desirable to set some guidelines for solar access ordinances that must be met before federal facilities will have to comply.

52. New State Ice Co. v. Liebmann, 285 U.S. 262, 311 (1932).

Senate to provide for a program of matching grants to states.[53] The grants would assist states in reviewing their laws affecting solar access and in identifying legal options for ensuring access to sunlight. The studies would be voluntary but would have to meet guidelines laid down by the U.S. Department of Energy.[54] Such a program could help states determine appropriate roles for themselves and their subdivisions. It could also help assure that solar access laws are well adapted to local needs and that unnecessary laws are not enacted. The hurried adoption of ill-considered legislation, such as that enacted in New Mexico,[55] would be worse than no solar access protection at all.

States, regions, counties, and municipalities may also need aid in implementing the legislation they adopt. Much of the financial burden of this could be in kind, that is, in technical services (such as aerial or satellite photographs) given free or on a less-than-fully reimbursed basis.

The feasibility of giving direct responsibilities for land use planning and controls (such as solar access protection) to higher levels of government has been summed up this way:

> There are no easy solutions to the complex problems raised by the transfer of land planning and development responsibilities to national and state governments . . . In theory, coordination and integration are reasonably easy . . . In practice, they are more than a little difficult to achieve.[56]

53. S. 3027, 95th Cong., 2d Sess. (1978).

54. Ibid. This bill is sponsored by Senator Gary Hart of Colorado, who recently stated the following opinion: "Legislative initiatives in this area have traditionally been the prerogative of state and local governments and should clearly remain so. The Federal government can, however, provide the crucial support needed by states and municipalities that seek to press forward and address these emerging problems." Letter from Senator Hart addressed to colleagues (28 April 1978).

55. 1977 N.M. Laws ch. 169 (*see* Chapter 9 for discussion; Appendix B for full text).

56. MANDELKER, ENVIRONMENTAL AND LAND CONTROLS LEGISLATION, p. 406.

13 Predicting and Resolving Disputes

Lawsuits are often the result of poorly conceived laws. Careful drafting of a solar access law can greatly reduce the possibility of future litigation. If a law or regulation clearly describes rights and duties, leaves little to the discretion of courts or administrators, and avoids provisions that are constitutionally questionable, it should leave few lawsuits in its wake. But because of the novelty of solar access protection, some conflicts are bound to arise. Communities using solar access zoning overlays or collector recordation systems may find their efforts challenged as going too far or not far enough. A community may have to defend its legislation both on its face and as it is applied to specific cases. In addition, conflicts are likely to arise between private parties when access to sunlight is blocked.

Among the various remedies to these conflicts are the rather unconventional ones of mediation and arbitration.[1] For reasons which follow, these remedies may be especially useful in conflicts over solar access. In addition, a recent survey found four major concerns of the American public with regard to our court system: delays in going to trial, the high cost of going to court, the high cost of

1. In both mediation and arbitration, the parties involved in a conflict agree to submit their dispute to a mutually agreed upon third party. In arbitration, the decisions of this third party (the arbitrator) are binding and final. The decision can be enforced just like a court judgment. Mediators can only try to guide the parties to areas where they can reach agreements. Mediators do not render "decisions" that are binding.

lawyers, and too few judges.[2] Mediation and arbitration may offer ways around some of these problems.

Potential Conflicts

Local efforts to protect solar access will be welcomed by both current and prospective solar collector owners. But individuals who live adjacent to solar collectors, neighborhood associations, and land developers may feel adversely affected by such legislation and may attempt to have it set aside.

The broadest challenge to a community's solar access law would be one that questions the validity of the ordinance on its face. Such a challenge could be predicated on several theories. First, opponents may contend that the state enabling legislation does not permit the locality to adopt, for example, a zoning plan or a recordation procedure to protect solar access. Second, they may claim that the ordinance fails to meet constitutionally imposed minimum standards of due process. The alleged defects could involve such things as the methods of providing notice, the type of hearing provided for, the decision-making process, takings of property without compensation, or the provision for judicial review of the decision. Third, opponents may charge that the new law sets up impermissible categories that treat similarly situated persons unequally. The criteria used to map various solar overlay zones or to shape solar envelopes or easements may be questioned.

These challenges are basic, since they attack either the power of the locality to enact solar access legislation or the very constitutionality of the legislation. Only one remedial route is open to resolve such a challenge: the judicial process. Nonjudicial options such as mediation, arbitration, or administrative hearings are inappropriate forums since none is able to result in a ruling on the power of the locality to act or the constitutionality of the legislation. The remedy available is likewise straightforward: declaratory relief that the local zoning plan or permit ordinance is or is not valid on its face, and an injunction against its operation if it is invalid.

More frequently, challenges to the way an access law is interpreted will occur. The validity of a specific decision made by a local authority applying the solar access legislation may be questioned. The local authority will often be a zoning board that determines

2. This survey was conducted by the opinion research firm of Yankelovich, Skelly and White, Inc., and reported in Stuart Auerbach and Laura A. Kiernan, *Public Want Court Reform, Poll Says,* Washington Post, 19 March 1978, p. A2.

whether to issue a building permit or to allow a zoning change, special permit, or a zoning variance. The losing party or someone else who disagrees with the decision may try to have it reversed or modified on the basis of its impact on solar access. These are some possible sources of conflict over the interpretation of solar access laws:

Collector owner vs. local authority. This type of dispute may be triggered by the denial of a solar collector recordation to a collector owner. Or the collector owner may challenge a decision that grants a neighbor or developer a development right that results in the solar collector being shaded. If the access law uses a nuisance approach, the collector owner may argue that the public prosecutor should sue (if the law does not give private persons standing) or that a nuisance does indeed exist.

Neighborhood association vs. local authority. This type of dispute may arise when the local authority grants a solar access permit to a collector owner over the objections of the neighborhood association, which may fear the solar energy system or the accompanying restriction will lower property values in the immediate area.

Adjacent landowner vs. local authority. There are several circumstances in which the neighbor of a solar user may challenge the decision of the local authority: when the neighbor is denied a permit to add to his structure; when he is ordered to trim vegetation shading a protected solar collector; or when a nearby solar collector owner is granted a solar access recordation over his opposition.

Developer vs. local authority. Developers may seek redress when the local authority denies a building permit or zoning variance, when a zoning variance is granted subject to certain conditions, or when a collector owner is granted a solar access recordation over the opposition of the developer.

In general, appellants can base their challenges to specific decisions on several grounds, any of which may form the basis of reversal: the decision did not comply with the mandate of the state enabling legislation; it exceeded the bounds of the "police power"; or the decisionmakers failed to observe the minimum standards of due process and equal protection.

Remedial Routes and Remedies

Once a conflict has arisen, there are several methods of dispute resolution, including: (1) nonjudicial approaches such as arbitration, mediation, or informal consideration; (2) administrative hearing, appeal, and judicial review; and (3) filing of a lawsuit in a court of general jurisdiction. Whichever remedial route a government elects to allow in solar access disputes, the law creating a right of action should:

- □ Give collector owners direct access to the remedy (rather than forcing them to wait for an official to initiate action on their behalf)
- □ Specify who pays the costs of the proceedings and lawyers fees
- □ Spell out a method for calculating damages
- □ Specify whether damages are acceptable in lieu of restoring solar access[3]

Mediation

Mediation is a technique often used to resolve disputes involving fairly small sums of money, such as conflicts between merchants and customers. Major disputes may also be submitted to mediation. The popularity of this method of settling arguments is growing.[4] Mediation may become more common if current legislation is enacted that would fund state programs to resolve minor disputes through mediation and arbitration.[5]

Using nonjudicial methods to settle a solar access dispute is likely to be confined to instances of conflicts between neighbors. (This technique is used in Japan to resolve exactly this kind of conflict.[6]) The mediator does not have to be appointed by the government; parties to the dispute can select their own.

3. *See* comments to remedies sections of the prototype legislation in Chapters 5 and 7.

4. *See,* e.g., MEDIATION/ARBITRATION: A NEW CONCEPT FOR CONFLICT RESOLUTION (Northbrook, Ill.: Soil Testing Services, Inc., 1977). Mediation is also being seriously considered as a way to resolve environmental disputes. *See* ENVIRONMENTAL MEDIATION: AN EFFECTIVE ALTERNATIVE? (Palo Alto, California: RESOLVE, Center for Environmental Conflict Resolution, 1978).

5. *See* Dispute Resolution Act, S. 957, 95th Cong., 1st Sess. (1977).

6. *See* Gail Feingold Takagi, *Designs on Sunshine: Solar Access in the United States and Japan,* CONNECTICUT LAW REVIEW 10 (1977): 123, 140. In Tokyo, for example, it is common for the developer, architect, and neighbors to discuss the effects of a proposed building on sunshine before an application for a building permit is filed. If the parties cannot resolve objections to the proposed construction, local government officials will attempt to mediate informally to help the parties reach a compromise. In addition, the Tokyo municipal government provides a Sunshine Problem Consultation Service to advise residents on resolving disputes among themselves. The service offers explanations of the laws and diagrams and models to illustrate

Mediation offers an appealing first-level attempt at resolving disputes. There is a great flexibility in remedies since the particular remedies applied may be whatever the parties agree to.[7] But because of the purely voluntary nature of mediation, its effectiveness is limited and back-up remedies will still be needed. Moreover, it does not lend itself to the broad category of solar access disputes where the legislation itself or the administrative decision of a local authority is challenged.

Arbitration

Like mediation, arbitration is becoming more widely used and encouraged. Chief Justice Warren E. Burger has said:

> As the work of the courts increases, delays and costs will rise and the well-developed forms of arbitration should have wider use. Lawyers, judges and social scientists of other countries cannot understand our failure to make greater use of the arbitration process to settle disputes.[8]

Arbitration has long been viewed as a good method of settling commercial and construction industry disputes and, more recently, disputes between labor and management as well. Its proponents claim that arbitration is successful because:

- Disputes are resolved promptly and inexpensively
- Arbitration proceedings are flexible and informal
- Decisionmakers are experts in the field
- The award is final

their effect. There is also a municipally sponsored Committee for Adjustment of Sunshine Disputes that is composed of experts who help resolve disputes through consultations held in markets, department stores, and other public places. Ibid.

7. For instance, an environmental dispute over the impacts of a shopping center was successfully mediated in Montgomery County, Maryland. The "remedies" were interesting—citizen support was granted in exchange for lower buildings, special landscaping, and a ban on drive-in restaurants and gas stations. Margot Hornblower, *Mediation Studied as Way to Resolve Environmental Disputes,* Washington Post, 13 January 1978, p. A7. Flexible remedies could be extremely useful in resolving solar access conflicts. Landowner A, for example, could agree to let Landowner B share excess heat from the solar system attached to A's structure, in exchange for the right to partially shade B's collectors. This is a type of remedy that would not be available in a court.

8. *Agenda for 2000 A.D.: Need for Systematic Anticipation* (address delivered at the National Conference on the Causes of Popular Dissatisfaction with the Administration of Justice, St. Paul, Minn., 7 April 1978) (text included in MEDIATION/ARBITRATION, p. 11).

Because of its record of success, arbitration has been considered as a possible method of resolving solar access disputes that arise between neighbors (for example, when one person's structure or vegetation blocks another person's protected access to sunlight).[9] In such instances, the resolution of these disputes could be achieved by a voluntary agreement to arbitrate or by statutorily mandated arbitration.[10] Both voluntary and mandatory arbitration raise legal and practical questions, the answers to which will decide whether arbitration is a feasible method for resolving solar access disputes.

Voluntary arbitration. Parties to an agreement can sometimes anticipate potential problems and agree to submit future disputes to binding arbitration. They can also agree in advance on the governing principles that will apply should a conflict occur, and on who the arbitrators will be. Some states will not enforce awards made under agreements to submit future disputes to arbitration. In those states, solar collector owners will have to reach arbitration agreements after a dispute arises. If the parties agree to submit their dispute after it arises, every state except Oklahoma will enforce an arbitration award. Where there is a general interest in solar energy use, however, as in a solar subdivision situation, prior agreements are more likely. In this situation, persons buying into the subdivision could agree to be bound by arbitration.

If the disputing parties agree to be bound by AAA rules, the following procedure is likely: The AAA would refer the dispute to a single arbitrator or a panel of arbitrators. The cost of such an arbitration would include administrative fees, probably according to the AAA's Commercial Arbitration Rules, which are based on the amount in controversy. (Arbitrators serve voluntarily, without cost, unless the case is prolonged; then, the parties agree to the payment of a reasonable fee.) For public policy purposes, a community could decide to pay arbitration fees in solar access cases in lieu of having to pay court costs.

The arbitrator, following rules and guidelines of the AAA, is in charge of the arbitration. He or she decides on matters such as admissibility of evidence, the use of affidavits and subpoenas, and the

9. Oklahoma is considering a bill that would provide for compulsory arbitration of disputes over interference with solar access. Okla. H.B. 1741 (1978).

10. It is possible to use a procedure that falls somewhere between mediation and arbitration. Parties could give up the right to formal procedures, but allow the third party more "muscle" than in mediation to use only if encouragement and negotiation fail.

scope of investigation conducted outside the hearing room. The tone of the arbitration is usually informal. Any evidence that seems relevant is admitted. Expert testimony may be presented in any format, including lecture or discussion style. Expert testimony will be crucial to solar access dispute resolution, and may be more readily grasped in this informal atmosphere than in the adversary cross-examination of a court room. The use of attorneys in an arbitration is not always necessary but is not prohibited. The AAA seems to encourage the use of attorneys.

The scope of the arbitrator's power to make awards is broad:

> The arbitrator may grant any remedy or relief which he deems just and equitable and within the scope of the agreement of the parties, including, but not limited to, specific performance of a contract.[11]

Arbitrators can grant money damages, specific performance, and injunctive relief in the form of stopping a person from continuing to do something that amounts to a breach of contract.[12] Thus, in a solar access dispute the arbitrator might award a person whose solar collector was blocked by his neighbor's tree an award of money damages to compensate for the loss of sunlight or order the neighbor to trim or remove the tree. In the event the neighbor had not yet planted the tree, the arbitrator might prohibit the neighbor from planting a tree where it would shade the collector. An award of punitive damages, however, is likely to be struck down as being outside the scope of the arbitrator's authority.[13] But punitive damages are not an essential ingredient of a good solar access law.

By voluntarily submitting to arbitration, the parties agree in advance that the award will be final. Appeals to the judicial system from such awards will be granted only on very narrow grounds: (1) the arbitrator's conduct was clearly partial; (2) a potential basis for bias existed and was not disclosed by the arbitrator in advance; or (3) the arbitrator exceeded his or her authority in the conduct of the hearing.

It has been estimated that the average commercial arbitration takes 60 days or less.[14] The average case handled by the AAA takes

11. Rule 42 of the American Arbitration Association's commercial arbitration rules.

12. American Arbitration Association, A Manual for Commercial Arbitrators (New York, n.d.), p. 17.

13. *See* Garrity v. Lyle Stuart, Inc., 40 N.Y.2d 354, 353 N.E.2d 793, 386 N.Y.S.2d 831 (1976).

14. Martin Domke, Commercial Arbitration (Englewood Cliffs, N.J.: Prentice-Hall, 1965), p. 8.

about four months from filing to disposition. Arbitration is almost always less time-consuming than litigation.

Compulsory arbitration. Some states have decreed that certain classes of cases, generally those below a stated dollar amount, can be submitted to arbitration without the voluntary consent of the parties. Two statutes offer examples of how states could place solar access disputes in a category that would go to arbitration.

Since 1951, Pennsylvania has allowed each county Court of Common Pleas to require, by court rule, that all cases below a certain amount go first to an arbitration panel composed of three lawyers.[15] In general, the amount in controversy cannot exceed $10,000 in larger counties and $5,000 in smaller counties; cases involving title to real estate are excluded. In addition to its compulsory feature, the statute also permits the court to send cases to arbitration where the parties agree to submit their dispute voluntarily. The state decrees that the award made by the "trial by lawyer" is given the effect of a final judgment. The procedures followed in compulsory arbitration are similar to those of voluntary arbitration, but with significant differences. The major difference is the appeal process. The Pennsylvania law provides that either party has the right to demand a trial *de novo,* provided a nonrecoverable fee is paid to the court. The purpose of this appeal provision is to overcome a constitutional objection that mandatory arbitration denies the parties a right to trial by jury and that it denies due process of law.

Since 1970, New York has had a statute that permits a judicial administrative board to promulgate rules for the compulsory arbitration of claims not exceeding $6,000.[16] Under this law, pilot projects were begun in the courts of Monroe and Bronx counties. As under the Pennsylvania law, the case is referred by the court to a panel of volunteer lawyers who are paid $45 each by the state. Besides its compulsory provision, the law permits parties to submit to arbitration voluntarily. The law provides for appeal in the form of a trial *de novo,* upon payment of the arbitrator's fee to the court.

Are compulsory arbitration laws successful? Do they provide a model for resolving solar access disputes? There are no clear answers to these questions. In both Pennsylvania and New York, the compulsory arbitration laws have transferred many thousands of cases from courtrooms to the arbitrators' hearing rooms. Even with the appeal

15. PA. STAT. ANN. tit. 5, § 30, as amended (Purdon).
16. N.Y. JUD. LAW § 213(10) (McKinney Supp. 1977).

procedure, the demand for trials *de novo* has been small and, where appeals have been taken, most decisions have been upheld.[17] As a result of the reduced backlog of cases, the processing of cases remaining in the courts has been accelerated, with time before trial going from fifteen months to two.[18] The substitution of arbitration for litigation probably reduces the cost of resolving an individual case and, it could be argued, of providing justice.

On the other hand, these advantages of speed, economy, and finality depend upon compliance with the arbitrator's award and minimizing the number of appeals. The historically high rate of compliance with arbitration decisions has been based on voluntary submission to arbitration; compulsory arbitration may result in a lower rate of compliance and in added litigation if the winner must go to court to enforce an arbitration award.

Conclusion. Arbitration is generally speedier and less expensive than litigation. It follows procedures that are less formal and forbidding than trial and offers the parties more flexibility and convenience in scheduling a hearing. For these reasons, the resolution of solar access disputes by arbitration presents an appealing alternative to courtroom litigation. Arbitration cannot be used, however, when questions are raised about the validity of the law itself or when an administrative decision is challenged.

Administrative Procedures

Just as the careful drafting of a law will lessen the number of lawsuits brought because of it, so will the careful planning of the administrative procedures used to decide solar access issues reduce litigation. In some cases, a state solar access law will set out the procedures to be followed; the community must then ensure compliance with them. In other situations, each locality will be responsible for both developing and carrying out the procedures.

Whether a community selects zoning overlays or recordation to protect solar access, administrative hearings are the method most often used to determine that protection. In general, administrative hearings will be held to:

- ☐ Establish regulations regarding solar access zoning or recordation

17. *See* National Institute for Consumer Justice, Staff Studies on Business Sponsored Mechanisms for Redress and Arbitration (n.p., n.d.), pp. 76–79.

18. Ibid.

- □ Determine whether solar access protection should be granted or denied in specific cases
- □ Review the correctness of a decision to grant or deny protection in specific cases

If the procedures that a community follows for its hearings are suspect, lawsuits and other challenges can be expected. Preplanning should concentrate on at least three major areas:

- □ Procedures to provide sufficient notice of solar access hearings
- □ Procedures to ensure that the hearings are fair and productive
- □ Procedures to ensure an adequate appellate process

How is notice to be given? A community using zoning or solar collector recordations to protect solar access must design its administrative hearing process to guarantee full participation to all parties. Adequate notice is a vital step to ensure that all who are affected are notified of a possible action in time to file an objection. Notice can take the form of certified letters, signs posted on the property, or publication in a newspaper of general circulation (posting and publication alone may be inadequate). (Notice is not an issue in arbitration or mediation since one must agree to participate in those proceedings.)

What kinds of hearings are required? Zoning is traditionally thought of as legislative action, which ordinarily does not require a hearing prior to the decision. Most state zoning enabling statutes, however, provide for a hearing, though the precise requirements vary. The process of deciding whether to grant a recordation will be similar to decisions to grant or deny permits of various sorts. Decisions on permits are almost universally delegated to an administrator. Much litigation has been spawned over the question of how a hearing must be conducted to accord due process. Hearings are relatively formal: evidence is received and arguments are made by the parties involved. Definite issues of law or fact are tried. No jury is used. Hearings may terminate in final orders. This initial decision may then be reviewed by the relevant agency. (Both pieces of prototype legislation in this report provide for a hearing procedure, integrating such elements as notice requirements and burden of proof into the community's existing administrative process.)

How formal will a hearing be? A community's administrative procedure should specify the following: how formal or informal the hearing will be; the rights of each side to examine witnesses; what evidence can be presented; where the burden of proof lies on particular issues; that the hearing is open to the public; and that the hearing be conducted on the record.

There should be some standard for the decisionmaker to follow in reaching a final decision. Courts have fluctuated on the issue of how much specificity is required to uphold an administrative decision, but the more explanation that is given to justify a particular decision, the more likely courts are to uphold it on appeal.

How can an administrative decision be appealed? After the initial administrative decision, the losing party must sometimes appeal to an administrative board of appeals before his case will be accepted by the community's court of general jurisdiction. The local ordinance should spell out the correct procedure to appeal a decision. Courts require all administrative remedies to be exhausted before they will consider a matter. Too many steps before reaching the final determination of an access conflict would be expensive to all concerned, including the public.

Conclusion. If the established administrative procedures are fair and not too burdensome, they can circumvent the need for some types of litigation. Solar access laws should specify administrative remedies where they are appropriate.

Judicial Approach

Adjudication is the only possible forum for some types of conflicts, such as constitutional questions, appeals from administrators' or arbitrators' decisions, or when an injunction is sought.[19] Solar

19. An injunction will sometimes be the only satisfactory remedy in a solar access conflict. People may often decide to install collectors for noneconomic reasons, such as reduced dependence on utility companies (suggesting either a fear of future fuel shortages, higher prices, or a desire to be self-sufficient). REGIONAL AND URBAN PLANNING IMPLEMENTATION, INC., FEDERAL INCENTIVES FOR SOLAR HOMES: AN ASSESSMENT OF PROGRAM OPTIONS, final report to the U.S. Department of Housing and Urban Development (Cambridge, Mass., July 1977), p. II-12. ELI is indebted to Rene Dubos for mailing us information on a recent French decision where the Court of Saint-Nazaire ordered a 15-story building destroyed. The owner of a home nearby had sued for reparation for interference with enjoyment of his property. The tall building apparently also violated urban regulations. *Tribunaux: La Cour D'Appel de Rennes Ordonne La Demolition d'un Tour de Quatorze Etages a la Baule,* PRESSE-ENVIRONNEMENT, 28 April 1978, p. 282. Such a severe remedy is available only through courts.

collector owners may appeal a decision denying a solar access permit or granting a zoning exemption to someone who will shade their collector; other landowners, developers, or neighborhood associations might appeal decisions granting a solar collector recordation or denying a zoning exemption.

Sometimes a court will be the first forum for a particular solar access dispute. For example, a lawsuit is necessary to challenge the validity of a statute: that the community's permit ordinance or overlay zoning plan does not comport with state enabling legislation, for example, or is unconstitutional. Conflicts between neighbors and private nuisance suits[20] will typically be handled in the courts because procedures for mediation and arbitration seldom are established.[21]

Judges or hearing examiners have many remedies from which to choose, although they may be more restrained in ordering some types of compromises than are arbitrators. Persons who challenge the validity of a solar access law on its face would seek declaratory relief and an injunction prohibiting the application of the law. The losing party who appeals an administrative decision would seek a reversal of the administrative order granting or denying the permit or zoning exemption. A solar collector owner who sues a neighboring landowner in a civil court for shading collectors could seek money damages, an injunction ordering the breaching party to stop any activity that shades the collector impermissibly, specific performance (if a private agreement is involved), or an order to trim vegetation. In a private suit based on nuisance law, a court may award the plaintiff either damages or an injunction.

If a solar access law declares violations to be misdemeanors, conflicts will be resolved in criminal proceedings. Such proceedings seem less suited to solar access disputes than are civil proceedings, as courts would be restricted to imposing fines or jail sentences. Fines paid to a local government by a violator would not directly compen-

20. ELI strongly recommends against allowing the traditional nuisance remedy of self-help. Allowing collector owners to go on neighbors' lands to cut their vegetation would breed litigation.

21. One commentator suggests that requiring a sensitive user (the owner of a shaded collector) to act as a plaintiff will deter him from seeking a remedy. He might fear the expense, as well as the difficulties of meeting the burden of proof. It is suggested that this problem could be overcome if nuisance actions are made administrative rather than judicial at the initial levels. A zoning board or like body could then handle complaints at an informal meeting. James Atwood, *An Economic Analysis of Land Use Conflicts,* STANFORD LAW REVIEW 21 (January 1969): 315.

sate the collector owner. Jailing a violator would be no more satisfactory. However, injunctions are not available as part of the criminal process to stop "crimes" before they occur. Criminal proceedings suffer from another drawback: A collector owner may have to wait for the public prosecutor to sue on his or her behalf.

Conclusion. Judicial actions will experience both costs and delays. The typical sequence of events can be complex. Both parties will have to hire attorneys to file complaints and responses. Even when an out-of-court settlement is possible, it often does not occur until the eve of trial, when great expenses have already been incurred. Where faster, less expensive remedies are possible, states and localities should consider requiring their use.

Appendices

Appendix A
More Solar Access Basics

This appendix is provided for the use of the technical staffs of legislators and officials, and for anyone else with the fortitude to probe technical aspects of solar access more deeply. An understanding of some of the complexities involved will help ensure that legislation accurately reflects local needs and conditions. Chapter 2, "Solar Access Basics," contains a generalized discussion of these issues. Although someone drafting a solar access law could work from the generalizations, a more detailed knowledge of how the generalizations were reached would be helpful in choosing standards concerning, for example, times of protection or envelope heights and slopes.

The Movements of the Sun

The relationship between time of day and the sun's position varies with the seasons. The sun reaches its peak altitude at solar noon at all times of year. The path is symmetrical about solar noon: the sun rises and sets the same length of time before and after solar noon, and during the day the sun's altitude is the same at equal lengths of time before and after solar noon. The sun's position at any particular time is generally specified by its altitude (or angle measured from the horizon) and its azimuth (or angle measured from true—not magnetic—south),[1] as shown in Figure A-1.

1. Azimuth is negative in the morning, zero at solar noon, and positive in the afternoon. For navigational purposes, azimuth is usually measured from true north, but the custom in the solar energy field is to measure from south. (Note: sometimes azimuth is treated as positive in the morning and negative in the afternoon; there is no universal custom.)

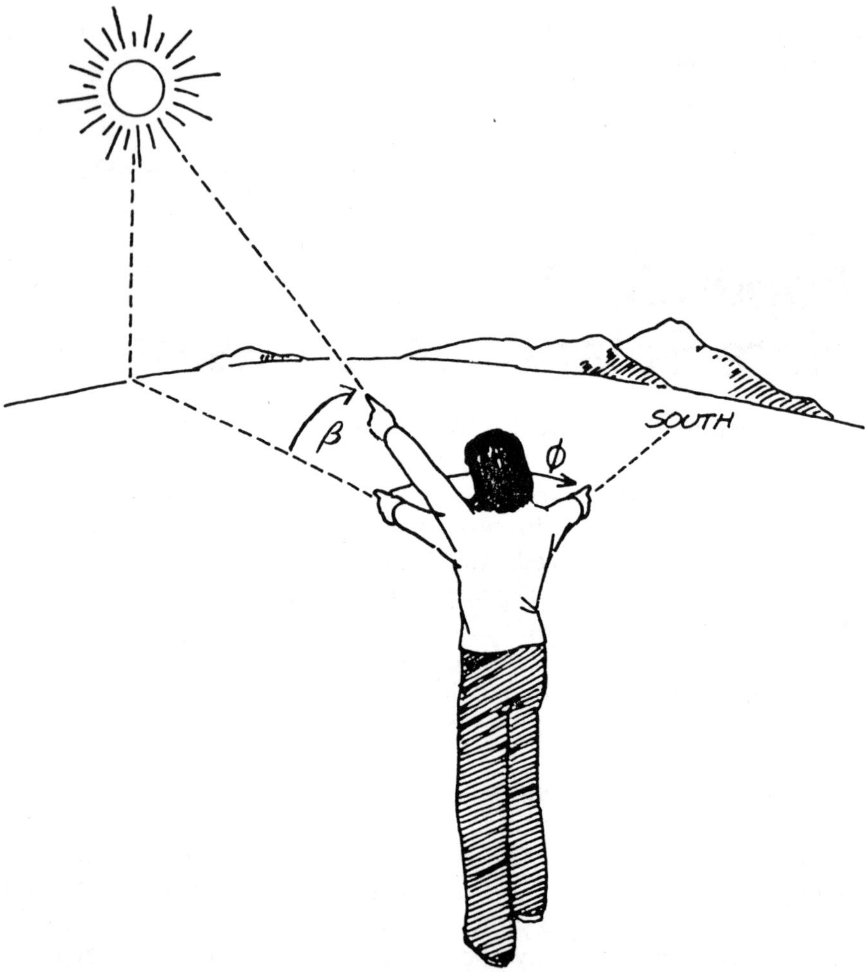

From *The Solar Home Book* by Bruce Anderson, Brick House Publishing Company, Andover, Mass. 01810, 1976. Reprinted by permission.

FIGURE A-1
DESCRIBING THE SUN'S POSITION

The solar altitude **β** is the angle between the sun and the horizon, and the azimuth **Φ** is measured from the true south (azimuth is positive to the west and negative to the east of south).

While these generalities are true, the specific position of the sun in relation to any particular observer depends on the latitude, time, and date. The higher the latitude, the lower the sun's altitude will be at any particular time and vice versa. The farther from noon the time is, the smaller the sun's altitude and the larger the magnitude of the azimuth will be. The closer the date is to the winter solstice, the lower the sun's altitude will be and the smaller the magnitude of the azimuth. The closer the date is to the summer solstice, the higher the sun's altitude will be and the larger the magnitude of the azimuth. Tables A-1 and A-2 illustrate these points. (Also see Figure 2-2 in Chapter 2.)

TABLE A-1
EFFECT OF LATITUDE
Winter Solstice (December 21)

Solar	32° N		40° N		48° N	
Time	Altitude	Azimuth	Altitude	Azimuth	Altitude	Azimuth
10 A.M.	27.6°	−31.2°	20.7°	−29.4°	13.6°	−28.2°
12 Noon	34.6°	0.0°	26.5°	0.0°	18.6°	0.0°
2 P.M.	27.6°	+31.2°	20.7°	+29.4°	13.6°	+28.2°

As one goes north, the altitude and magnitude of the azimuth fall.

TABLE A-2
EFFECT OF TIME
For 40° N, Winter Solstice (December 21)

Solar Time		Altitude	Azimuth
Sunrise	7:25 A.M.	0.0°	−58.7°
	8 A.M.	5.5°	−53.0°
	9 A.M.	14.0°	−41.9°
	10 A.M.	20.7°	−29.4°
	11 A.M.	25.0°	−15.2°
	12 Noon	26.5°	0.0°
	1 P.M.	25.0°	+15.2°
	2 P.M.	20.7°	+29.4°
	3 P.M.	14.0°	+41.9°
	4 P.M.	5.5°	+53.0°
Sunset	4:35 P.M.	0.0°	+58.7°

The sun is lower and farther from due south the further from noon it is. The altitude is the same for equal lengths of time from noon, but the signs differ (+ after noon; − before) as indicated in Table A-3.

TABLE A-3
EFFECT OF DATE
For 40° N

	Winter Solstice (December 21)		Equinoxes (Mar. 21/Sept. 21)		Summer Solstice (June 21)	
Solar Time	**Altitude**	**Azimuth**	**Altitude**	**Azimuth**	**Altitude**	**Azimuth**
4:00 A.M.	—	—	—	—	0 °*	−121.3°
5:00 A.M.	—	—	—	—	4.2°	−117.3°
6:00 A.M.	—	—	0 °†	− 90.0°	14.8°	−108.4°
7:00 A.M.	0 °‡	−58.7°	11.4°	−80.2°	26.0°	− 99.7°
8:00 A.M.	5.5°	−53.0°	22.5°	−69.6°	37.4°	− 90.7°
9:00 A.M.	14.0°	−41.9°	32.8°	−57.3°	−41.9°	− 80.2°
10:00 A.M.	20.7°	−29.4°	41.6°	−41.9°	59.8°	65.8°
11:00 A.M.	25.0°	−15.2°	47.7°	−22.6°	69.2°	41.2°
12:00 Noon	26.5°	0.0°	50.0°	0.0°	73.4°	0.0°

* Sunrise at 4:35 A.M. † Sunrise 6:00 A.M. ‡ Sunrise 7:25 A.M.

The altitudes are lower and the azimuths closer to south at the winter solstice; the altitudes are higher and the azimuths farther from south at the summer solstice.

Shadows and Shading

Formulas for computing shadow lengths to any desired degree of accuracy are at the end of this appendix. We summarize the implications of the formulas in tables and words. Table A-4 shows the effect of time of day on total shadow lengths.

TABLE A-4
TOTAL SHADOW LENGTHS IN FEET PER FOOT OF OBSTRUCTION
For 40° N

Solar Time	Winter Solstice	Equinoxes	Summer Solstice
12 Noon	2.0	0.8	0.3
11 A.M.	2.2	0.9	0.4
10 A.M.	2.7	1.1	0.6
9 A.M.	4.0	1.6	0.9
8 A.M.	10.4	2.4	1.3
7 A.M.	Sunrise: 7:25 A.M.	4.9	2.1
6 A.M.	—	Sunrise: 6:00 A.M.	3.8
5 A.M.	—	—	13.5
4 A.M.	—	—	Sunrise: 4:35 A.M.

As would be expected, shadow lengths for any particular time of day are shorter at the equinoxes and shortest at the summer solstice compared to the winter solstice.

Table A-5 shows the effect of ground slope on total shadow lengths (these particular times of day were chosen because they have

TABLE A-5
SHADOW LENGTHS IN FEET PER FOOT OF OBSTRUCTION FOR SELECTED SLOPES AND TIMES
For 40° N

	Solar Time	Level Ground	5% S* Slope	10% S Slope	5% N Slope	10% N Slope	5% W† Slope	10% W Slope	5% E Slope	10% E Slope
Winter	10 A.M.	2.7	2.4	2.2	3.0	3.4	2.8	3.0	2.5	2.3
Solstice	9 A.M.	4.0	3.5	3.1	4.7	5.7	4.7	5.5	3.5	3.2
	8 A.M.	10.4	7.9	6.4	15.2	27.9	17.8	61.7	7.4	5.7
Equinoxes	10 A.M.	1.1	1.1	1.0	1.2	1.2	1.2	1.2	1.1	1.0
	9 A.M.	1.6	1.5	1.4	1.6	1.7	1.7	1.8	1.5	1.4
	8 A.M.	2.4	2.3	2.2	2.5	2.6	2.7	3.1	2.2	2.0
	7 A.M.	4.9	4.7	4.6	5.2	5.4	6.5	9.6	4.0	3.3
Summer	9 A.M.	0.9	0.9	0.9	0.9	0.9	0.9	1.0	0.8	0.8
Solstice	8 A.M.	1.3	1.3	1.3	1.3	1.3	1.4	1.5	1.2	1.2
	7 A.M.	2.1	2.1	2.1	2.0	2.0	2.3	2.6	1.9	1.7

* "5% S" means slope is downward to the south at a rate of 5 feet per 100 feet (horizontal measure).

† Note, because morning sunlight slopes down toward the west and afternoon sunlight slopes down toward the east, to use the morning figures for the afternoon, you must use figures from the W columns for east slope and E columns for west slope. For example, at 2 P.M., winter solstice, 5% E slope, the correct shadow figure is 2.8, *not* 2.5.

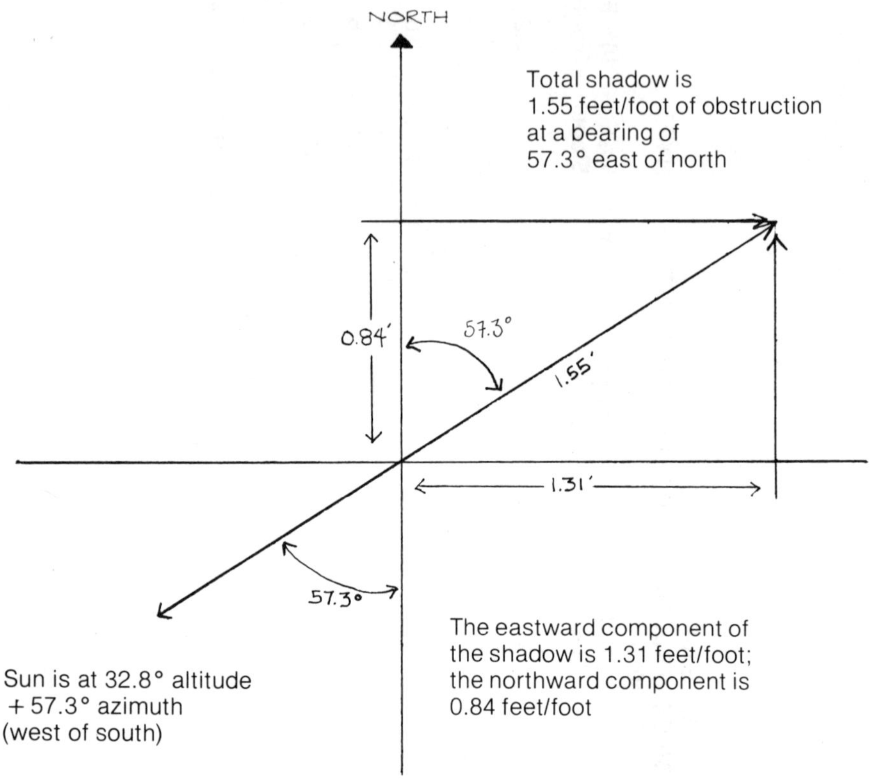

FIGURE A-2
NORTHWARD AND EASTWARD SHADOW COMPONENTS
Example is for level ground at 40°N, 3 P.M., on the equinoxes.

frequently been used in enacted or proposed solar access legislation). The effect of slope on total shadow length at the equinoxes and summer solstice is much less severe and shadows are in general much shorter than winter solstice shadows. The winter solstice shadows appear to be the main limiting factor on nearby development if solar access is to be protected.[2]

At the winter solstice, shadow lengths are extremely sensitive to north slopes in the morning and afternoon, to west slopes in the morning and east slopes in the afternoon. For example, for a 10 percent north slope, shadow lengths are more than 40 percent longer at

2. For winter solar space heating and year-round solar hot water. But for summer active solar air conditioning and photovoltaics, this may not be so; see Table A-6's eastward parts of shadows and accompanying text.

9 A.M. and 3 P.M. than on level ground. That means that the sphere of influence on permitted nearby development will be much larger for a north slope than for level ground if shadows on collectors are prohibited from 9 A.M. to 3 P.M. solar time. This effect will be even greater for latitudes farther north than 40°N, for slopes greater than 10 percent (which are developed, too), or for protecting hours beyond 9 A.M. to 3 P.M. A shadow has a direction in addition to its length. The direction of the shadow is opposite the sun's position, or, in other words, its bearing relative to north is the same as the sun's bearing relative to south—the azimuth (see Figure A-2).

Table A-6 breaks down shadows into their northward and eastward components.[3]

TABLE A-6
COMPONENTS OF SHADOW LENGTH IN FEET PER FOOT OF OBSTRUCTION
For 40° N on Level Ground

	Solar Time	Total Length	Bearing (Azimuth)	Northward Part*	Eastward Part†
Winter Solstice	8 A.M.	10.4	−53.0	6.3	−8.3
	9 A.M.	4.0	−41.9	3.0	−2.7
	10 A.M.	2.7	−29.4	2.3	−1.3
	2 P.M.	2.7	+29.4	2.3	+1.3
	3 P.M.	4.0	+41.9	3.0	+2.7
	4 P.M.	10.4	+53.0	6.3	+8.3
Equinoxes	7 A.M.	4.9	−80.2	0.8	−4.9
	8 A.M.	2.4	−69.6	0.8	−2.3
	9 A.M.	1.6	−57.3	0.8	−1.3
	10 A.M.	1.1	−41.9	0.8	−0.8
	2 P.M.	1.1	+41.9	0.8	+0.8
	3 P.M.	1.6	+57.3	0.8	+1.3
	4 P.M.	2.4	+69.6	0.8	+2.3
	5 P.M.	4.9	+80.2	0.8	+4.9
Summer Solstice	7 A.M.	2.1	−99.7	−0.3	−2.0
	8 A.M.	1.3	−90.7	0.0	−1.3
	9 A.M.	0.9	−80.2	0.1	−0.9
	3 P.M.	0.9	+80.2	0.1	+0.9
	4 P.M.	1.3	+90.7	0.0	+1.3
	5 P.M.	2.1	+99.7	−0.3	+2.0

* Southward when negative † Westward when negative

Several new insights can be drawn from this table. First, shading problems in the summer will not come from the south but from the east and west. Both the equinox and summer solstice figures show eastward and westward shadows that are generally longer than northward shadows. Second, the early morning and late afternoon

3. Figured using the formulas at the end of this appendix.

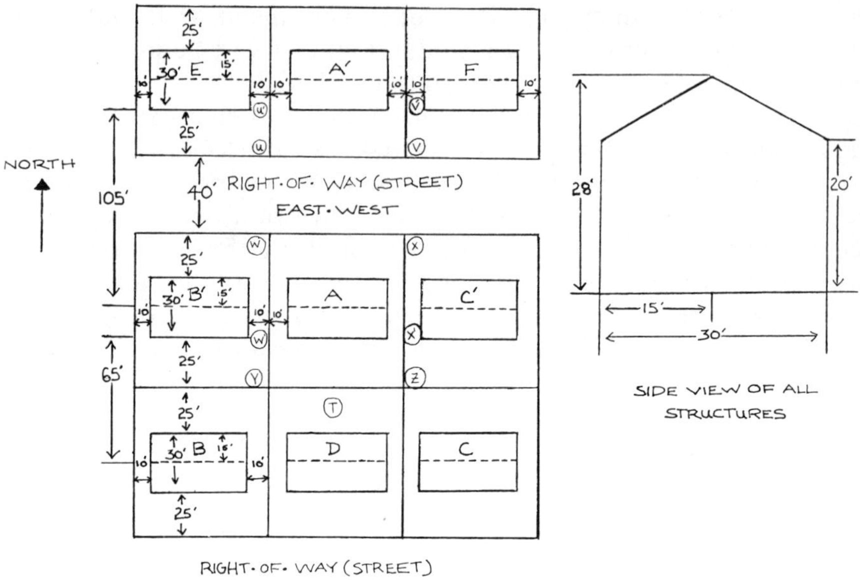

FIGURE A-3
RESIDENTIAL DEVELOPMENT EXAMPLE

40°N latitude. Actual building height, 28 feet to peak; zoned for 35 feet maximum height, setback 25 feet, rearyard 25 feet, and sideyard 10 feet. Buildings 50 by 30 feet (1,500 square feet) on 70 foot by 80 foot lots (5,600 square feet). 40 foot street right-of-way.

figures for the summer solstice indicate southward shadows, so that protection of active solar air conditioning or photovoltaic applications (when they become economic) may include protection against shading by properties to the north of the collector. This is an example of how changes in technology may call for changes in solar access laws.

The most significant point here is that the penetration of a property by a shadow from another property will be at a rate lesser than indicated by the total shadow figure if the shadow crosses the property line diagonally.[4] Using a plausible developmental situation (Fig-

4. Some commentators to the contrary notwithstanding. *See*, e.g., SANDY F. KRAEMER, SOLAR LAW, (Colorado Springs: Shepard's, Inc., 1978), p. 222 (left half of figure 36); Richard Stein, *Basic Considerations,* in A FORUM ON SOLAR ACCESS: PROCEEDINGS OF A FORUM ON SOLAR ACCESS HELD BY THE NEW YORK STATE LEGISLATIVE COMMISSION ON ENERGY SYSTEMS (Rockville, Md.: National Solar Heating and Cooling Information Center, 1977), p. 19 (where the spacing of buildings said to be needed is based on total shadow length rather than the northward part).

ure A-3), what can be said about the winter solstice solar access available to houses A and A′ (ignoring trees for the moment)? The structure shadow lengths are indicated in Table A-7.

TABLE A-7
STRUCTURE SHADOW LENGTHS IN FEET
For Figure A-3 Example on Winter Solstice

Solar Time		28′ Present Height	35′ Maximum Allowed
8 A.M.	Total	292	364
	North	176	220
	West	233	291
9 A.M.	Total	113	141
	North	84	105
	West	75	94
10 A.M.	Total	74	93
	North	65	81
	West	36	46

For house A, which is 65 feet north of the peaks of houses B and C, full south-wall access is available from 10 A.M. to 2 P.M. with the existing 28-foot structures; the shadow from B and C is 65 feet northward, not the full 74 feet of the total shadow length. But 7 feet of the south wall would be in shade if B or C were built up to the allowed 35 feet.[5] At 9 A.M. and 3 P.M., only 6 feet of A's south wall would be shaded for 28-foot high B and C (this would go up to 13 feet if B and C were 35 feet tall.) Existing access is rather good, and A and others similarly situated would benefit greatly if houses situated as B and C are were not allowed to be built any higher.

For house A′, which is 105 feet north of the peaks of houses B′ and C′, full south-wall access would be available from 9 A.M. to 3 P.M. even if B′ and C′ were built up to 35 feet; the shadow from B′ and C′ would be 105 feet, not the full 113 feet of the total shadow length. With B′ and C′ 28 feet tall, A′ has 21 feet of south yard in the sun at 9 A.M. and 3 P.M. This access would have to be counted as excellent and would still be good even if B′ and C′ built up to the maximum permitted height.

But what about vegetation? Although the existing zoning provides considerable protection against shading by buildings (with exceptions discussed in Chapter 4), trees can be a problem in such a development. Consider what modest-sized trees, say 30 feet high and 30 feet wide,[6] at circled locations U, V, W, X, Y, and Z might do to A or A′'s otherwise good south-wall access. The 30-foot trees will cast shadows at 9 A.M. and 3 P.M. extending about 90 feet north and

5. Figured using the formulas at the end of this appendix.

6. Many trees are, of course, much taller and wider than this.

about 81 feet west (at 9 A.M.) or east (at 3 P.M.). Tree U will shade about 22 feet up A''s south wall at 3 P.M. (this will include part of the roof as well, since the wall is 20 feet tall) and Tree V will do the same at 9 A.M. Trees Y and Z will do the same to A. Trees W and X, 65 feet from A''s south wall, will also shade the lower 8 feet of A''s south wall at 3 P.M. and 9 A.M. respectively. This all assumes that the trees are planted at the edge of the street right-of-way or at the B–B′ or C–C′ property lines. If the trees are larger or planted farther north, the loss of access will be more severe, perhaps even blocking most rooftop access, unless they are planted between the buildings in the side yards (at U′, V′, W′, or X″). Such plantings would help limit summer overheating (lowering air conditioning loads) from the east and west but not so severely limit winter access. For a new development, trees might be limited to such locations. In an existing development there will usually be trees along the street (often owned or controlled by the city) and trees elsewhere (as in the backyards) as well.[7] Note that with trees in this sort of residential situation, the shading can often come from the eastern or western neighbor, not just from the south.

Effect of Time System on Shadows

The specific effect on shadow lengths if standard time is used at different places within a particular time zone is indicated in Table A-8. Similar tables can be prepared for other latitudes, with similar results. If standard time is used west of the standard meridian, there is more morning and less afternoon protection. Neighbors to the east will be more severely limited in their permitted development than western neighbors. Furthermore, the solar collector user will be afforded inferior access compared to the use of solar time, since the sunlight is less intense, for example, between 8:30 A.M. and 9:00 A.M. solar time than between 2:30 P.M. and 3:00 P.M. solar time[8] (unless the locality has unusual afternoon overcast just after 2:30 P.M. solar time). This is also true in the case of standard time used east of the standard meridian, except that morning and afternoon, as well as east and west, are reversed.

7. A large enough tree at location T in D's back yard could block off noon access for A, even though shadows are only two feet per foot of obstruction at noon.

8. For various models and formulas for figuring available solar radiation *see* the discussion in T. KUSUDA AND D. ISHII, HOURLY SOLAR RADIATION DATA FOR VERTICAL AND HORIZONTAL SURFACES ON AVERAGE DAYS IN THE UNITED STATES AND CANADA: National Bureau of Standards Buildings Science Series, no. 96 (Washington, D.C.: U.S. Government Printing Office, April 1977), and their references.

TABLE A-8
EFFECT OF LONGITUDE ON SHADOWS IN FEET PER FOOT OF OBSTRUCTION
For 40° N on the Winter Solstice, Standard Time, on Level Ground, Ignoring the Equation of Time

	7½ °W of* Standard Meridian		3¾ °W of Standard Meridian		Solar Time and at Standard Meridian		3¾ °E of Standard Meridian		7½ °E of Standard Meridian	
	9 A.M.	3 P.M.	9 A.M.	3 P.M.	9 A.M.	3 P.M.	9 A.M.	3 P.M.	9 A.M.	3 P.M.
Total	5.7	3.2	4.7	3.5	4.0	4.0	3.5	4.7	3.2	5.7
Northward	3.9	2.6	3.3	2.7	3.0	3.0	2.7	3.3	2.6	3.9
Eastward†	−4.2	+1.9	−3.3	+2.2	−2.7	+2.7	−2.2	+3.3	−1.9	+4.2
Equivalent Solar Time	8:30 A.M.	2:30 P.M.	8:45 A.M.	2:45 P.M.	9:00 A.M.	3:00 P.M.	9:15 A.M.	3:15 P.M.	9:30 A.M.	3:30 P.M.

* Some areas are further east or west of standard meridian than the cited 7½ ° distance. For example, the Eastern Standard Time Zone extends to 12½ ° west, the Central Standard Time Zone to 14 ° west, and the Mountain Standard Time Zone to 13 ° west of the standard meridian.

† Westward when negative (morning)

TABLE A-10
EFFECT OF EQUATION OF TIME ON SHADOWS IN FEET PER FOOT OF OBSTRUCTION
For 40° N on Level Ground

	November 21			December 21			January 21		
	Standard 9 A.M.	Solar 9 A.M., 3 P.M.	Standard 3 P.M.	Standard 9 A.M.	Solar 9 A.M., 3 P.M.	Standard 3 P.M.	Standard 9 A.M.	Solar 9 A.M., 3 P.M.	Standard 3 P.M.
Total	3.64	3.29	3.01	3.96	4.02	4.09	2.96	3.29	3.73
Northward	2.52	2.37	2.25	2.96	2.99	3.02	2.23	2.37	2.55
Eastward	−2.63	±2.29	+2.01	−2.63	±2.69	+2.75	−1.95	±2.29	+2.72
Equivalent Solar Time	9:13:48	9, 3	3:13:48	9:01:36	9, 3	3:01:36	8:48:48	9, 3	2:48:48

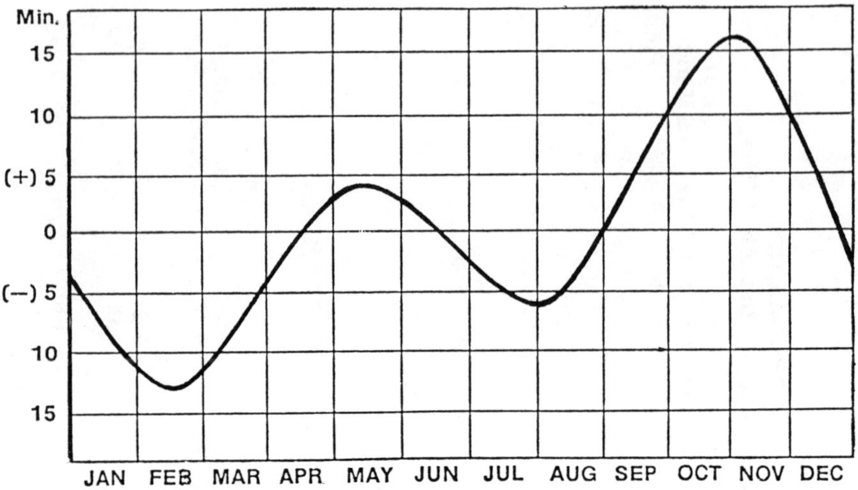

Source: Libbey-Owens-Ford, *Designing With the LOF Sun Angle Calculator* (1975), p. 22. Used with permission.

FIGURE A-4
EQUATION OF TIME

The "equation of time" effect is due to the variations in the length of the actual solar day compared to the average twenty-four hour day (the so-called mean solar day). This factor varies somewhat from year to year but is a function that is roughly as shown in Figure A-4. During the course of a year, the equation of time varies over a range of about half an hour. Widely used standard figures for the equation of time are shown in Table A-9.

TABLE A-9
EQUATION OF TIME IN MINUTES FOR THE 21st OF EACH MONTH

Solar Time Compared to Standard Time is:	Month	Equation of Time
LATE	January	– 11.2
	February	– 13.9
	March	– 7.5
EARLY	April	+ 1.1
	May	+ 3.3
LATE	June	– 1.4
	July	– 6.2
	August	– 2.4
EARLY	September	+ 7.5
	October	+ 15.4
	November	+ 13.8
	December	+ 1.6

Source: ASHRAE, *Applications Handbook* (1978), p. 58.2

The variation in shadow lengths is not profound, although noticeable when given to three significant figures as indicated in Table A-10.

Since the November and January shadow lengths, even after taking the equation of time into account, are still shorter than the December shadows,[9] no compensation for monthly variation in the equation of time appears justified. On the winter solstice there is some slight asymmetry in shading between morning and afternoon. This could justify an additional small correction of about two minutes to Table 2-1 of Chapter 2, as shown in Table A-11.

TABLE A-11
FULLY COMPENSATED STANDARD TIME
9 A.M. to 3 P.M. Solar Time Becomes:

Standard Time Hours	Relationship to Standard Meridian
9:32 A.M.–3:32 P.M.	7½°W
9:17 A.M.–3:17 P.M.	3¾°W
9:02 A.M.–3:02 P.M.	0°
8:47 A.M.–2:47 P.M.	3¾°E
8:32 A.M.–2:32 P.M.	7½°E

Using Planes to Define Solar Access

The development situation in Figure A-3 is reprinted in simplified form here as Figure A-5 to show ways of defining solar access in terms of space. As indicated earlier, in this situation A has full south-wall access from 10 A.M. to 2 P.M. and upper south-wall access from 9 A.M. to 10 A.M. and 2 P.M. to 3 P.M. (over six feet is in shade at 9 A.M. and 3 P.M.). How can one define angles for protecting A's access, either for an agreement with A's neighbors B, B′, C, C′, and D or for a public solar access law? The northward shadow length ratio (feet/foot of obstruction) for 40°N on the winter solstice is about 2.3 at 10 A.M. and 2 P.M. and about 3.0 at 9 A.M. and 3 P.M. (from Table A-6; the figures are 2.31 and 2.99 to three places). These represent slopes of 1/2.3 = .43 and 1/3.0 = .33 or angles of 23½° and 18½°.[10] Thus a specification that defined a plane sloping upward to the south beginning at the base of A's south wall with an angle of 23½° would define the space that the neighbors must stay within to assure A's full south-wall access from 10 A.M. to 2 P.M. Or the angle

9. This is also true for other months, latitudes, and times of day (10 A.M. to 2 P.M. and 8 A.M. to 4 P.M.) that ELI checked, but did not present in tables here.

10. These angles are figured from the arctangent of the reciprocal of the shadow length ratio: for example arctan (1/2.3) = 23½ deg.

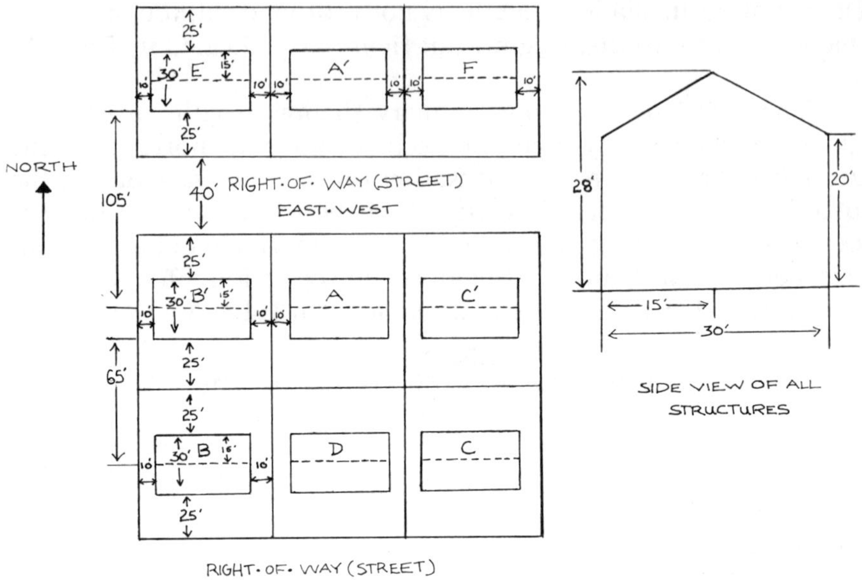

FIGURE A-5
RESIDENTIAL DEVELOPMENT EXAMPLE
Simplified from Figure A-3.

could be defined by the slope instead of degrees (unless the law requires degrees): with a slope of .43 or with a rise of 3 feet for each 7 feet from A's south wall (3/7 = .43). This last form of specification, rise to distance ratio, is probably the easiest to understand, measure, and enforce: an angle of 23½° could not be measured directly without a surveyor's transit. An alternative specification for A's upper south-wall access from 9 A.M. to 3 P.M. could specify that the plane begin 6′3″ up from the base of the south wall and be at an angle of 18½° or at a slope of .33 (or a rise of 1 foot for each 3 feet from the wall). These two planes are illustrated in Figures A-6 and A-7.

In these cases the plane was defined in relation to the property line at the northern edge of lots B, C, and D. The plane for full south-wall access would begin at B, C, and D's northern property line at 10′8″ and slope up, as before, at 23½°, 3/7, or .43. The plane for upper south-wall access would begin at 14′8″ and slope up, as before, at 18½°, 1/3, or .33. These planes are illustrated in Figures A-8 and A-9. The planes in Figures A-6 and A-8 are equivalent, as are the planes in Figures A-7 and A-9.

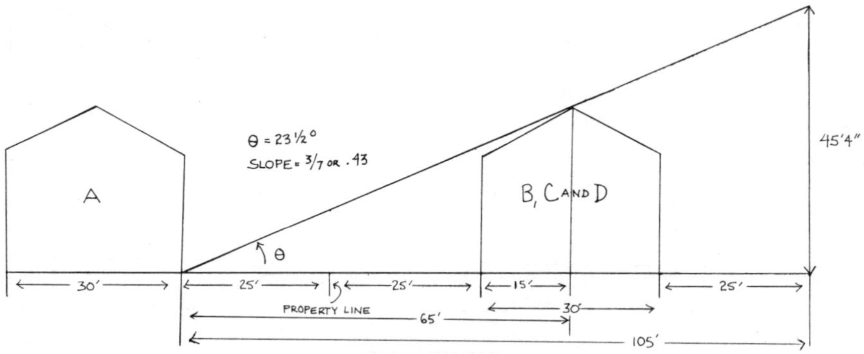

FIGURE A-6
COLLECTOR-BASED PLANE FOR FULL SOUTH WALL ACCESS
This is for 10 A.M. to 2 P.M. in example at 40°N.

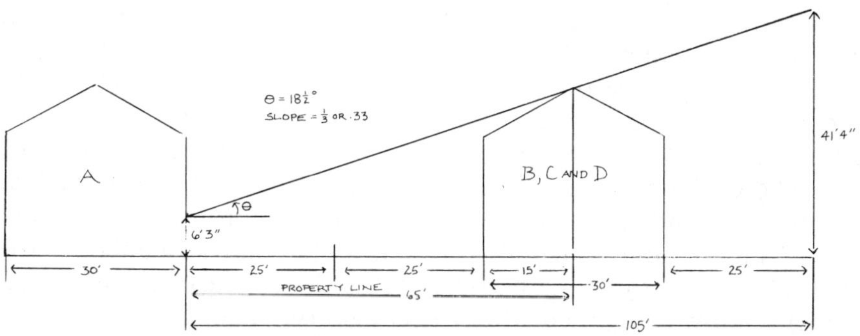

FIGURE A-7
COLLECTOR-BASED PLANE FOR UPPER SOUTH WALL ACCESS
This is for 9 A.M. to 3 P.M. in example at 40°N.

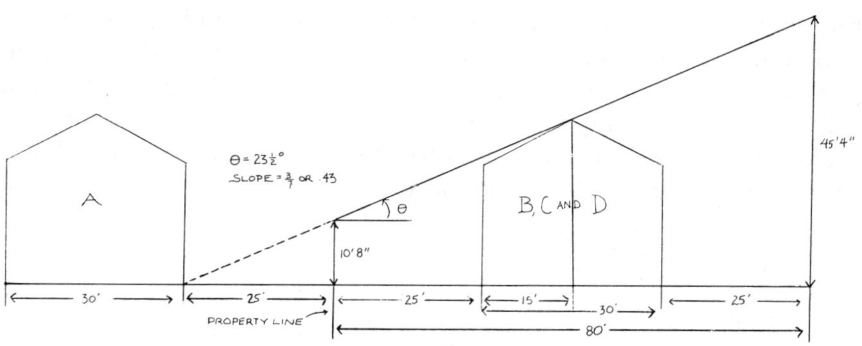

FIGURE A-8
PROPERTY-LINE-BASED PLANE FOR FULL SOUTH WALL ACCESS
This is for 10 A.M. to 2 P.M. in example at 40°N.

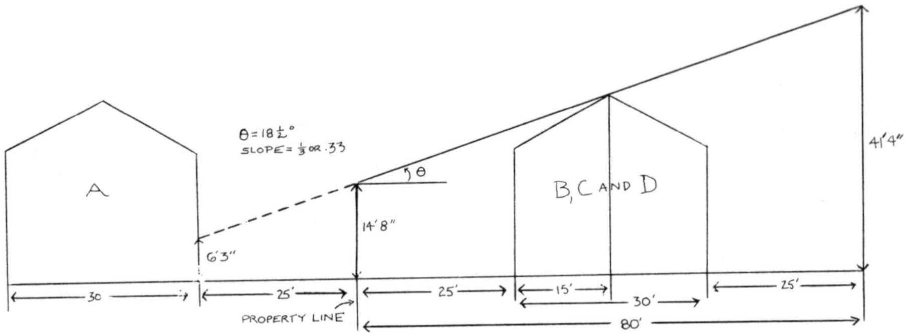

FIGURE A-9
PROPERTY-LINE-BASED PLANE FOR UPPER SOUTH WALL ACCESS
This is for 9 A.M. to 3 P.M. in example at 40°N.

In the same way, planes can be defined for rooftop access with the slope beginning at the lower edge of the roof (20 feet up from the base of the south wall). The angle (or slope) of the plane determines the hours of full access at the winter solstice (10 A.M. to 2 P.M. for 23½°, 9 A.M. to 3 P.M. for 18½°). (See Figures A-10 and A-11.)

These planes have all been based on the northward shadow components and slope upward to the south to protect against shadows from the south (including from B to the southwest and C to the southeast). The same method can be used with eastward (or westward) shadow components to protect against shading from the west (or east). This could be done in the kind of situation pictured in Figure A-5 to protect A against shading from trees on the lots of B′ or C′, the west and east neighbors (trees at points Y and Z of Figure A-5, for example). Again, these planes have been based on a simple, but plausible, residential development on level ground with the street and long axes of buildings running east and west, for protection against shading at the winter solstice. Any of these factors and others could be changed and the analysis be redone using the formulas at the end of this appendix to compute solar access and planes to specify various levels of solar access.

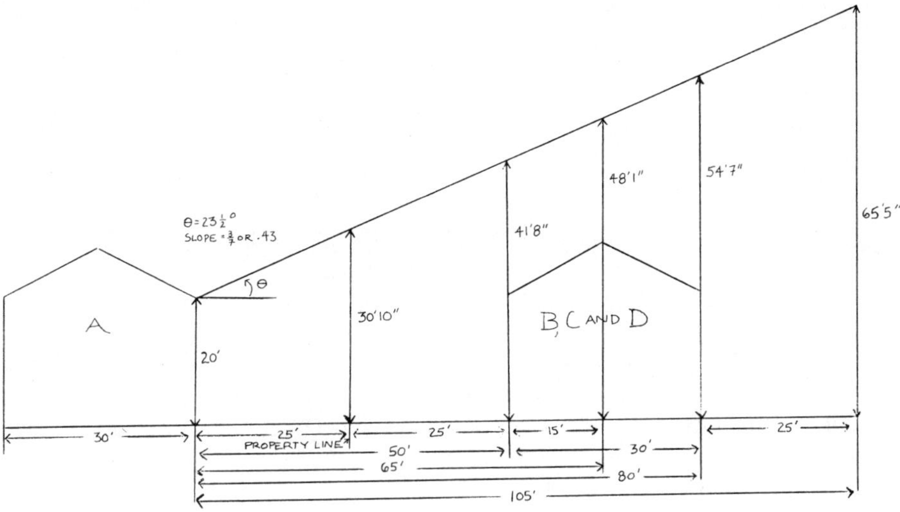

FIGURE A-10
COLLECTOR-BASED PLANE FOR ROOFTOP ACCESS
This is for 10 A.M. to 2 P.M. in example at 40°N.

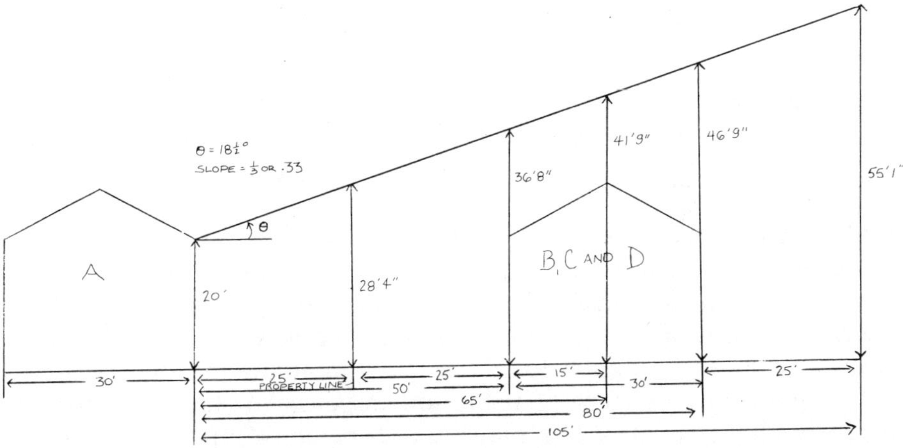

FIGURE A-11
ANOTHER COLLECTOR-BASED PLANE FOR ROOFTOP ACCESS
This is for 9 A.M. to 3 P.M. in example at 40°N.

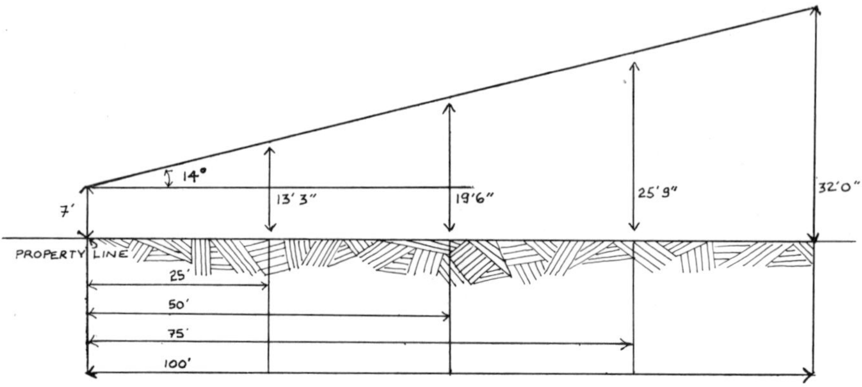

FIGURE A-12
IMPLIED PLANE OF PROTECTION FOR HYPOTHETICAL WALL

Example is for 7′ high wall and protection 8 A.M. to 4 P.M. on the winter solstice at 35°N.

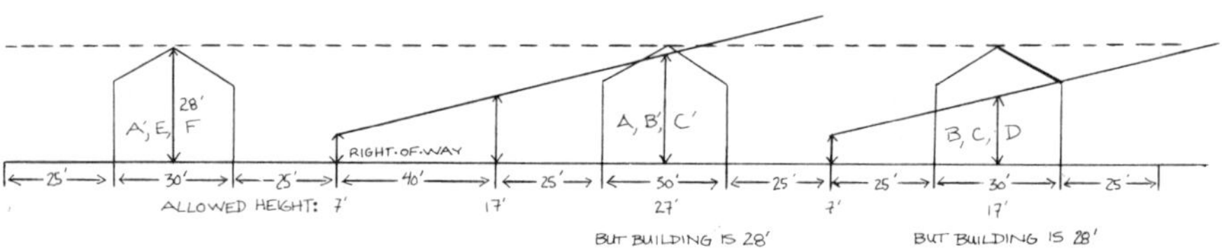

FIGURE A-13
INCOMPATIBILITY OF 7′, 14° HYPOTHETICAL WALL WITH EARLIER RESIDENTIAL DEVELOPMENT EXAMPLE

This is the same example as in Figure A-3.

Implications of a Hypothetical Wall Standard

Eisenstadt, Long, and Utton's study for the State of New Mexico suggests (but does not actually recommend) protection from 8 A.M. to 4 P.M. mountain standard time at the winter solstice for a hypothetical wall 7′ high at 35°N. latitude (Albuquerque, New Mexico).[11] Ignoring the effect of using uncompensated standard time, this is roughly equivalent to a plane at a slope of 14° up to the south beginning at an east–west property line at 7′ off the ground. This is illustrated in Figure A-12. This would provide a plane much too low for the kind of small lot development considered earlier. The incompatibility can be seen in Figure A-13. In short, the buildings in this

11. Melvin M. Eisenstadt, Stephen C. M. Long, and Albert E. Utton, *A Proposed Solar Zoning Ordinance,* mimeographed, prepared for the New Mexico Energy Institute (ERB-76-103, November 1977), p. 7, revised version, same title, URBAN LAW ANNUAL 15 (1978), p. 221.

situation are not tall enough to be compatible with this specific standard, but if the wall were taller or the hours less long, a hypothetical wall standard could be applied to such a development.[12]

Formulas for Calculating Sun Angles and Shadows

Sun angles

$\beta = \arcsin(\cos L \times \cos\delta \times \cos H + \sin L \times \sin\delta)$

$\Phi = \mathrm{sgn}H \times \arccos((\sin L \times \sin\beta - \sin\delta) / (\cos L \times \cos\beta))$

$\theta = \arccos(\cos\beta \times \cos(\Phi - \psi) \times \sin\Sigma + \sin\beta \times \cos\Sigma)$

where

β = solar altitude

Φ = solar azimuth (relative to south)

θ = angle of incidence between direct solar rays and the normal to the collector surface

L = latitude

δ = declination (angle between sun and plane of Earth's equator for the date in question)

H = hour angle (minutes from solar noon ÷ 4; negative in morning, positive in afternoon)

ψ = collector azimuth angle (measured from south)

Σ = collector tilt

Shadow length of vertical pole on sloping ground

TL (Total Length) = height / ($\tan\beta + \cos\gamma' \times \tan\Sigma'$)

Northerly component = TL $\times \cos\Phi$

Easterly component = TL $\times \sin\Phi$

where

$\gamma' = \Phi - \psi'$

ψ' and Σ' are azimuth and slope of ground

These are measured horizontally.

For $\tan\Sigma'$, just use slope (rise/horizontal running distance) if slope is a ratio rather than degrees.

To get figures to measure along sloping surface, divide by $\cos(\arctan(\cos\gamma' \times \tan\Sigma'))$.

12. This discussion shows only the implications of the hypothetical wall on east-west property lines. A similar analysis of the effect on north-south property lines shows a comparable incompatability with the residential development discussed.

Appendix B
Enacted Solar Access Laws

NOTE: Selected local and model solar access ordinances are reprinted in APA's *Guidebook for Planning Officials,* Appendix III. For reasons explained in the text, the Environmental Law Institute does not recommend that other jurisdictions enact laws based on the New Mexico or California laws.

New Mexico Solar Rights Act 1977 N.M. Laws Chapter 169

Relating to solar energy; providing declarations and findings concerning solar rights; providing a declaration of solar rights

Be it enacted by the Legislature of the State of New Mexico:

Section 1. Short Title—This act may be cited as the "Solar Rights Act."

Section 2. Declaration and Findings—The legislature declares that the state of New Mexico recognizes that economic benefits can be derived for the people of the state from the use of solar energy. Operations, research, experimentation and development in the field of solar energy use shall therefore be encouraged. While recognizing the value of research and development of solar energy use techniques and devices by governmental agencies, the legislature finds and declares that the actual construction and use of solar devices, whether at public or private expense, is properly a commercial activity which the law should encourage to be carried out, whenever practicable, by private enterprise.

Section 3. Definitions—As used in the Solar Rights Act:

A. "solar collector" means any device or combination of devices or elements which rely upon sunshine as an energy source, and which are capable of collecting not less than twenty-five thousand Btu's on a clear winter solstice day. The term also includes any substance or device which collects solar energy for use in:

(1) the heating or cooling of a structure or building;
(2) the heating or pumping of water;
(3) industrial, commercial, or agricultural processes; or
(4) the generation of electricity.

A solar collector may be used for purposes in addition to the collection of solar energy. These uses include, but are not limited to, serving as a structural member or part of a roof of a building or structure and serving as a window or wall; and

B. "solar right" means a right to an unobstructed line-of-sight path from a solar collector to the sun, which permits radiation from the sun to impinge directly on the solar collector.

Section 4. Declaration of Solar Rights—

A. The legislature declares that the right to use the natural resource of solar energy is a property right, the exercise of which is to be encouraged and regulated by the laws of this state. Such property right shall be known as a solar right.

B. The following concepts shall be applicable to the regulation of disputes over the use of solar energy where practicable:

(1) "beneficial use." Beneficial use shall be the basis, the measure and the limit of the solar right, except as otherwise provided by written contract. If the amount of solar energy which a solar collector user can beneficially use varies with the season of the year, then the extent of the solar right shall vary likewise;

(2) "prior appropriation." In disputes involving solar rights, priority in time shall have the better right except that the state and its political subdivisions may legislate, or ordain that a solar collector user has a solar right even though a structure or building located on neighborhood property blocks the sunshine from the proposed solar collector site. Nothing in this paragraph shall be construed to diminish in any way the right of eminent domain of the state of any of its political subdivisions or any other entity that currently has such a right; and

(3) "transferability." Solar rights shall be freely transferable within the bounds of such regulation as the legislature may impose. The transfer of a solar right shall be recorded in accordance with Chapter 71, Article 2, NMSA 1953.

C. Unless singular overriding state concerns occur which significantly affect the health and welfare of the citizens of this state, permit systems for the use and application of solar energy shall reside with county and municipal zoning authorities.

Section 5. Prior Rights Unaffected—Nothing in the Solar Rights Act shall be construed to alter, amend, deny, impair or modify any solar right, lease, easement, or contract right which was vested prior to the effective date of the Solar Rights Act.

Section 6. Effective Date—This act shall become effective on July 1, 1978.

California Solar Shade Control Act 1978 Cal. Stats. Chapter 1366

The people of the State of California do enact as follows:

Section 1. Chapter 12 (commencing with Section 25980) is added to Division 15 of the Public Resources Code, to read:

Chapter 12. Solar Shade Control

25980. This chapter shall be known and may be cited as the Solar Shade Control Act. It is the policy of the state to promote all feasible means of energy conservation and all feasible uses of alternative energy supply sources. In particular, the state encourages the planting and maintenance of trees and shrubs to create shading, moderate outdoor temperatures, and provide various economic and aesthetic benefits. However, there are certain situations in which the need for widespread use of alternative energy devices, such as solar collectors, requires specific and limited controls on trees and shrubs.

25981. As used in this chapter, "solar collector" means a fixed device, structure, or part of a device or structure, which is used primarily to transform solar energy into thermal, chemical, or electrical energy. The solar collector shall be used as part of a system which makes use of solar energy for any or all of the following purposes: (1) water heating, (2) space heating or cooling, and (3) power generation.

25982. After January 1, 1979, no person owning, or in control of a property shall allow a tree or shrub to be placed, or, if placed, to grow on such property, subsequent to the installation of a solar collector on the property of another so as to cast a shadow greater than 10 percent of the collector absorption area upon that solar collector surface on the property of another at any one time between the hours of 10 A.M. and 2 P.M., local standard time; provided, that this section shall not apply to specific trees and shrubs which at the time of installation of a solar collector or during the remainder of that annual solar cycle cast a shadow upon that solar collector. For the purposes of this chapter, the location of a solar collector is required to comply with the local building and setback regulations, and to be set back not less than five feet from the property line, and no less than 10 feet above the ground. A collector may be less than 10 feet in height, only if in addition to the five feet setback, the collector is set back three times the amount lowered.

25983. Every person who maintains any tree or shrub or permits any tree or shrub to be maintained in violation of Section 25982 upon property owned by such person and every person leasing the property of another who maintains any tree or shrub or permits any tree or shrub to be maintained in violation of Section 25982 after reasonable notice in writing from a district attorney or city attorney or prosecuting attorney, to remove or alter the tree or shrub so that there is no longer a violation of Section 25982, has been served upon such person, is guilty of a public nuisance as defined in Sections 370 and 371 of the Penal Code and in Section 3480 of the Civil Code. For the purposes of this chapter, a violation is hereby deemed an infraction. The complainant shall establish to the satisfaction of the prosecutor that the violation has occurred prior to the prosecutor's duty to issue the abatement notice. For the purpose of this section, "reasonable notice" means 30 days from receipt of such notice. Upon expiration of the 30-day period, the complainant shall file an affidavit with the prosecutor alleging that the nuisance has not been abated if the complainant wishes to proceed with the action. The existence of such violation of each and every day after the service of such notice shall be deemed a separate and distinct offense, and it is hereby made the duty of the district attorney, or the city attorney of any city the charter of which imposes the duty upon the city attorney to prosecute state infractions, to prosecute all persons guilty of violating this section by continuous prosecutions until the violation is corrected. Each and every violation of this section shall be punishable by a fine not to exceed five hundred dollars ($500).

25984. Nothing in this chapter shall apply to trees planted, grown, or harvested on timberland as defined in Section 4526 or on land devoted to the production of commercial agricultural crops. Nothing in this chapter shall apply to the replacement of a tree or shrub which had been growing prior to the installation of a solar collector and which, subsequent to the installation of such solar collector, dies.

25985. Any city, or for unincorporated areas, any county, may adopt, by majority vote of the governing body, an ordinance exempting their jurisdiction from the provisions of this chapter. The adoption of such an ordinance shall not be subject to the provisions of the California Environmental Quality Act (commencing with Section 21000).

25986. Any person who plans a passive or natural solar heating system or cooling system or heating and cooling system which would impact on an adjacent active solar system may seek equitable relief in a court of competent jurisdiction to exempt such system from the provisions of this chapter. The court may grant such an exemption based on a finding that the passive or natural system would provide a demonstrably greater net energy savings than the active system which would be impacted.

Sec. 2. Notwithstanding Section 2231 of the Revenue and Taxation Code, there shall be no reimbursement pursuant to that section nor shall there be any appropriation made by this act because the Legislature recognizes that during any legislative session a variety of changes to laws relating to crimes and infractions may cause both increased and decreased costs to local government entities and school districts which, in the aggregate, do not result in significant indentifiable cost changes.

Appendix C Solar Heating Systems

Solar Heating and Hot Water Systems: Functional Description

The basic function of a solar heating and domestic hot water system is the collection and conversion of solar radiation into usable energy. This is accomplished — in general terms — in the following manner. Solar radiation is absorbed by a collector, placed in storage as required, with or without the use of a transport medium, and distributed to point of use. The performance of each operation is maintained by automatic or manual controls. An auxiliary energy system is usually available for operation, both to supplement the output provided by the solar system and to provide for the total energy demand should the solar system become inoperable.

The conversion of solar radiation to thermal energy and the use of this energy to meet all or part of a dwelling's heating and domestic hot water requirements has been the primary application of solar energy in buildings.

The parts of a solar system — collector, storage, distribution, transport, controls and auxiliary energy — may vary widely in design, operation, and performance. They may, in fact, be one and the same element (a south-facing masonry wall can be seen as a collector, although a relatively inefficient one, which stores and then radiates or "distributes" heat directly to the building interior). They may also be arranged in numerous combinations dependent on function, component compatibility, climatic conditions, required performance, site characteristics, and architectural requirements.

Of the numerous concepts presently being developed for the collection of solar radiation, the relatively simple flat-plate collector has the widest application. It consists first of an absorber plate, usually made of metal coated black to increase absorption of the sun's energy. The plate is then insulated on its underside and covered with a transparent cover plate to trap heat within the collector and reduce convective losses from the absorber. The captured heat is removed from the absorber by means of a working fluid, generally air or water. The fluid is heated as it passes through or near the absorber plate and then transported to points of use, or to storage, depending on energy demand.

Reprinted from U.S. Department of Housing and Urban Development, Intermediate Minimum Property Standards Supplement: Solar Heating and Domestic Hot Water Systems (HUD: Washington, D.C., 1977), 5: § 4930.2; also available in pamphlet form from its authors, Massdesign Architects and Planners, Inc., as Solar Heating Primer (Cambridge, Mass.: AIA Research Corporation, n.d.).

The storage of thermal energy is the second item of importance since there will be an energy demand during the evening, or on sunless days when solar collection cannot occur. Heat is stored when the energy delivered by the sun and captured by the collector exceeds the demand at the point of use. The storage element may be as simple as a masonry floor that stores and then re-radiates captured heat, or as relatively complex as a latent heat storage. In some cases, it is necessary to transfer heat from the collector to storage by means of a heat exchanger (primarily in systems with a liquid working fluid). In other cases, transfer is made by direct contact of the working fluid with the storage medium (i.e., heated air passing through a rock pile).

The distribution component receives energy from the collector or storage, and dispenses it at points of use. Within a building, heat is usually distributed in the form of warm air or warm water.

The controls of a solar system perform the sensing, evaluation and response functions required to operate the system in the desired mode. For example, if the collector temperature is sufficiently higher than storage temperature, the controls can cause the working fluid in storage to circulate in the collector and accumulate solar heat.

An auxiliary energy system provides the supply of energy when stored energy is depleted due to severe weather or clouds. The auxiliary system, using conventional fuels such as oil, gas, electricity, or wood provides the required heat until solar energy is available again.

The organization of components into solar heating and domestic hot water systems has led to two general characterizations of solar systems: active and passive. The terms active and passive solar systems have not yet developed universally accepted meanings. However, each classification possesses characteristics that are distinctively different from each other. These differences significantly influence solar dwelling and system design.

An active solar system can be characterized as one in which an energy resource — in addition to solar — is used for the transfer of thermal energy. This additional energy, generated on or off the site, is required for pumps, blowers, or other heat transfer medium moving devices for system operation. Generally, the collection, storage, and distribution of thermal energy is achieved by moving a transfer medium throughout the system with the assistance of pumping power.

A passive solar system, on the other hand, can be characterized as one where solar energy alone is used for the transfer of thermal energy. Energy other than solar is not required for pumps, blowers, or other heat transfer medium moving devices for system operation. The major component in a passive solar system generally utilizes some form of thermal capacitance, where heat is collected, stored, and distributed to the building without additional pumping power. Collection, storage, and distribution is achieved by natural heat transfer phenomena employing convection, radiation, conduction, in conjunction with the use of thermal capacitance as a heat flow control mechanism.

Solar Heating and Hot Water Systems: Operational Description

A. Solar Heating System

Solar systems may be designed to operate in a number of different ways depending on function, required performance, climatic conditions, component and system design, and architectural requirements. Usually, however, solar systems are designed to operate in four modes. In a very basic manner, the four modes of solar system operation for both active and passive systems are described and illustrated below.

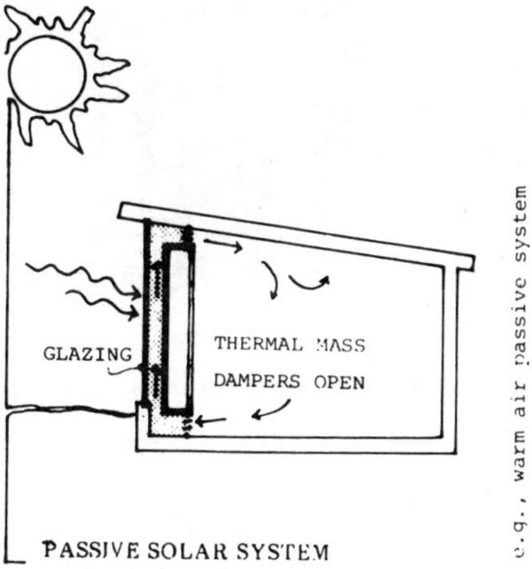

PASSIVE SOLAR SYSTEM

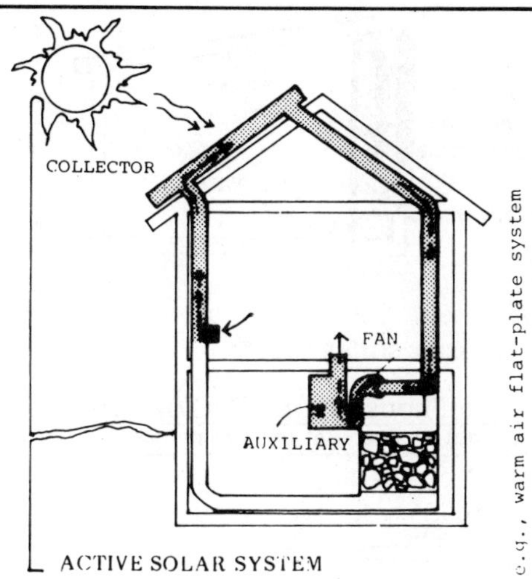

ACTIVE SOLAR SYSTEM

1. HEATING HOUSE FROM COLLECTOR Solar radiation captured by the collectors and converted to thermal energy can be used to directly heat the house.

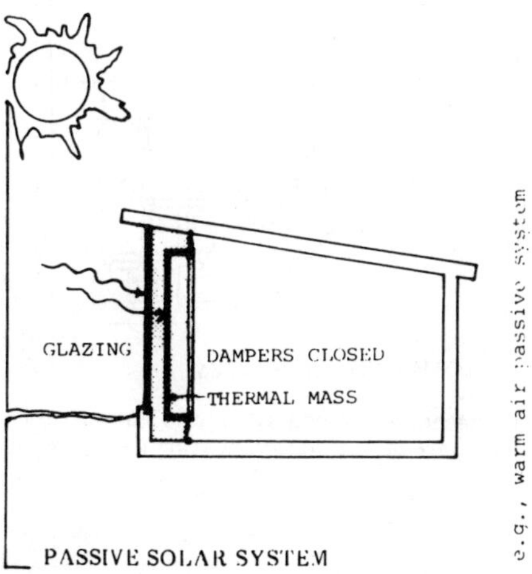

PASSIVE SOLAR SYSTEM

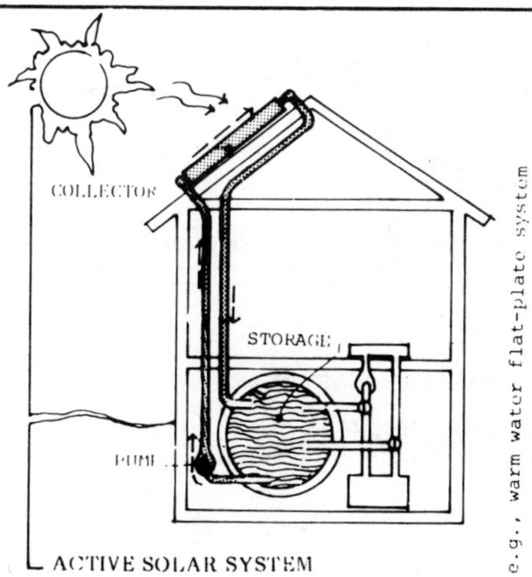

ACTIVE SOLAR SYSTEM

2. HEATING STORAGE FROM COLLECTOR If the house does not require heat, the captured (collected) thermal energy can be placed in storage for later use.

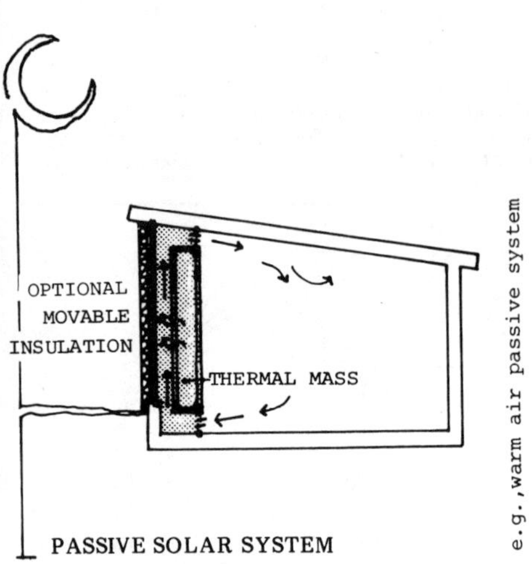

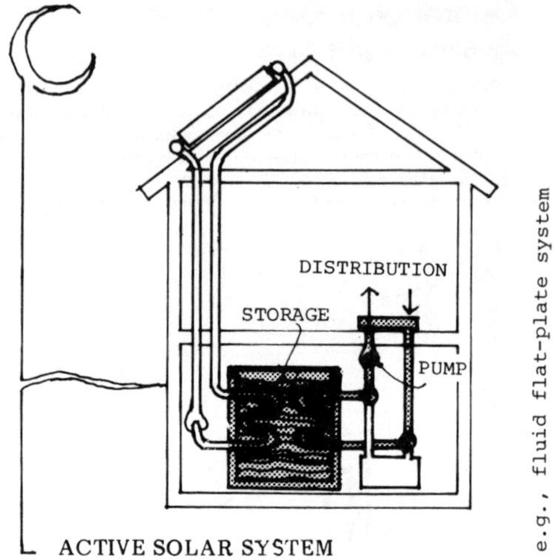

3. <u>HEATING HOUSE FROM STORAGE</u> Heat from storage can be removed to heat the house when the sun is not shining — at night or on consecutive sunless days.

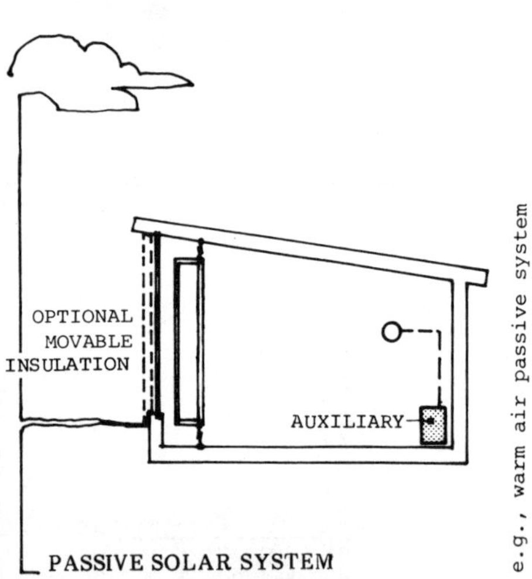

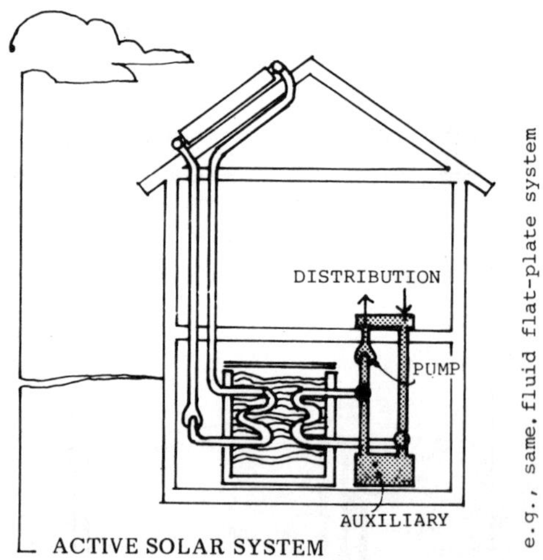

4. <u>HEATING HOUSE FROM AUXILIARY</u> If heat from the collector and storage is not sufficient to totally heat the house, an auxiliary system supplies all or part of the house's heating requirement.

B. Domestic Hot Water System

The solar system may also be designed to preheat water from the incoming water supply prior to passage through a conventional water heater. The domestic hot water preheat system can be combined with the solar heating system or designed as a separate system. Both situations are illustrated below.

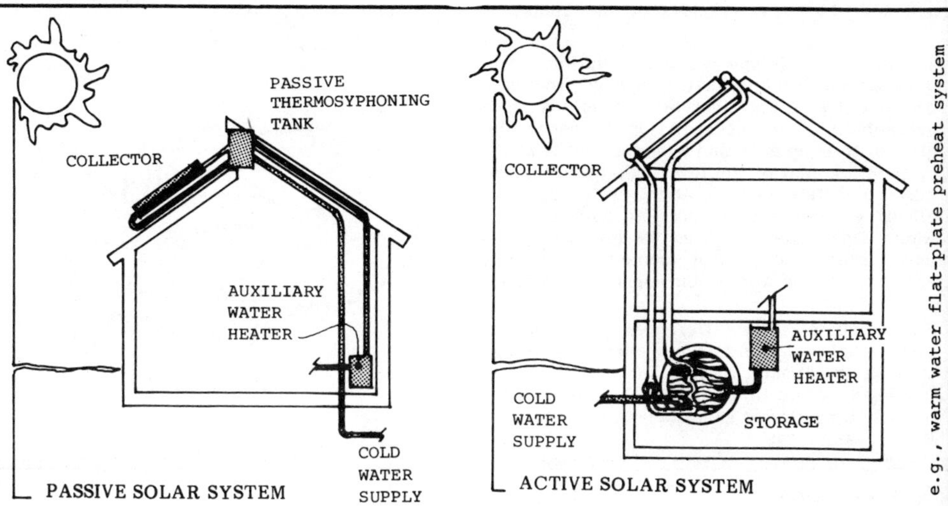

1. DOMESTIC HOT WATER PREHEATING - SEPARATE SYSTEM Domestic hot water preheating may be the only solar system included in some designs. A passive thermosyphoning arrangement is shown above.

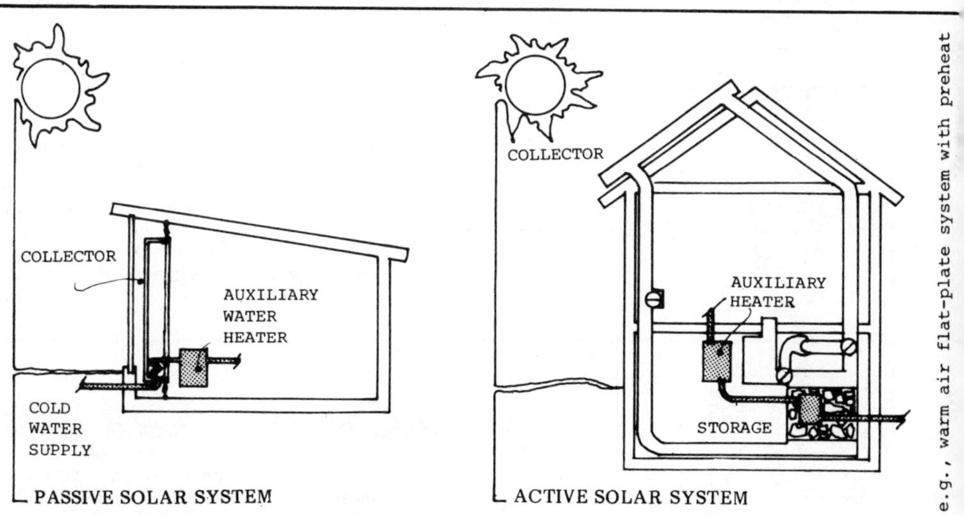

2. DOMESTIC HOT WATER PREHEATING - COMBINED SYSTEM Domestic hot water is preheated as it passes through heat storage enroute to the conventional water heater.

Basics of Solar Utilization

A. Climate

Solar radiation, wind, temperature, humidity and many other factors shape the climate of the United States. Basic to using solar energy for space heating and domestic hot water heating is understanding the relationship of solar radiation, climate and dwelling design.

The amount and type of solar radiation varies between and within climatic regions: from hot-dry climates where clear skies enable a large percentage of direct radiation to reach the ground, to temperate and humid climates where up to 40 percent of the total radiation received may be diffuse sky radiation, reflected from clouds and atmospheric dust, to cool climates where snow reflection from the low winter sun may result in a greater amount of incident radiation than in warmer but cloudier climates.

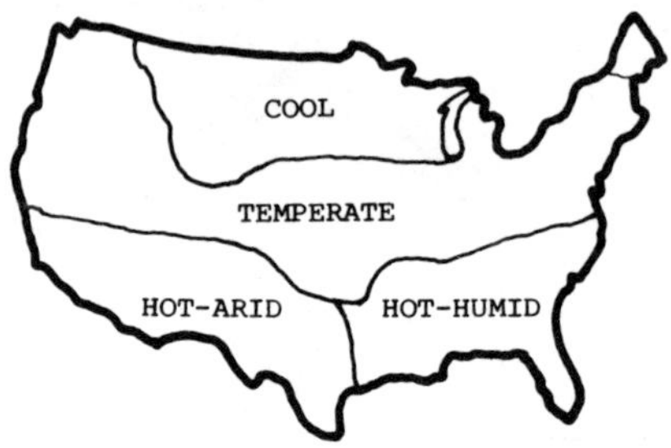

As a result of these differences in the amount and type of radiation reaching a building site, as well as in climate, season and application — heating or domestic hot water — the need for and the design of solar system components will vary in each locale.

B. Solar Radiation

The sun provides almost all of the earth's energy in the form of radiation. Solar energy, also known as solar radiation reaches the earth's surface in two ways : by direct (parallel) rays, and by diffuse (nonparallel) sky radiation. The solar radiation reaching a building includes not only direct and diffuse but also radiation reflected from adjacent ground and building surfaces. It is these three sources of solar radiation that may be used for space and domestic hot water heating.

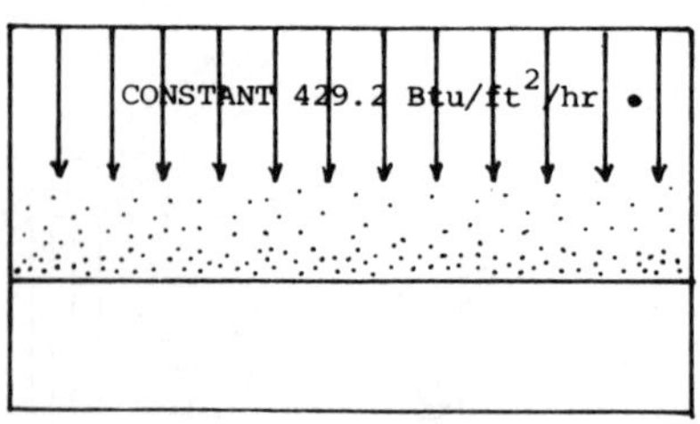

1. <u>THE SOLAR CONSTANT</u> A nearly constant amount of solar energy strikes the outer atmosphere = 429.2 BTU per square foot per hour. This quantity is known as the solar constant. A large amount of this energy, however, is lost in the earth's atmosphere,and cannot be regained regardless of collector design.

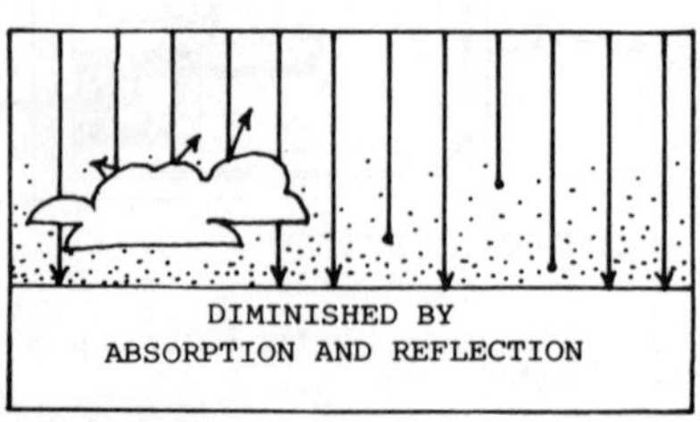

2. <u>ABSORPTION AND REFLECTION</u> On the average, almost half of the solar radiation reaching the earth's outer atmosphere is lost by absorption in the atmosphere and by reflection from clouds, as it passes through the atmosphere to the earth's surface. The radiation lost actually varies between 60% in Seattle, Washington to only 30% in Albuquerque, New Mexico.

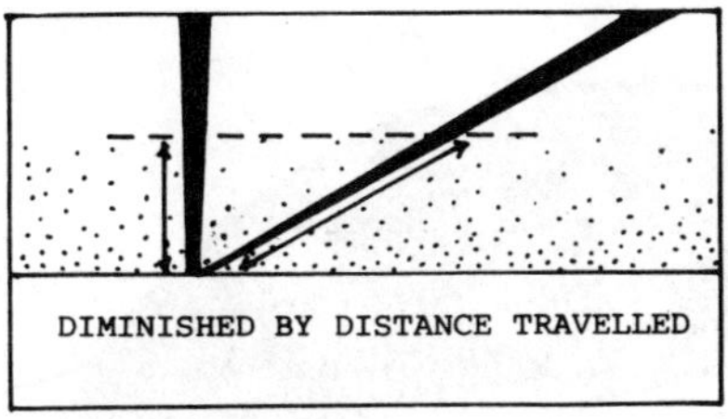

3. <u>EARTH'S ATMOSPHERE</u> As already state the radiation reaching the earth's surface diminished by the condition of the earth atmosphere: its vapor, dust and smoke conten At lower sun angles, the length of trav through the atmosphere is greatly increased, the relative amount of radiation received further diminished.

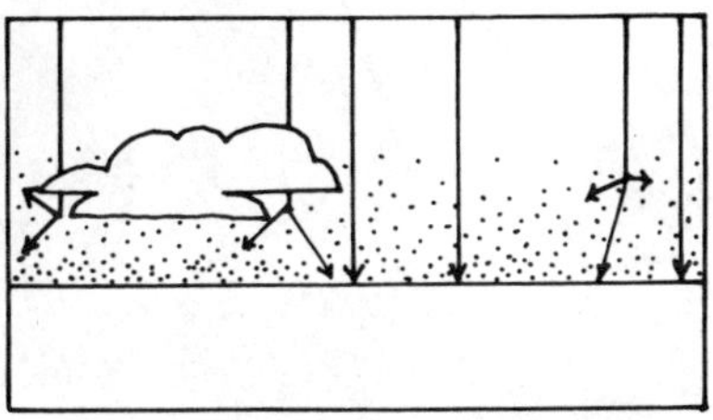

4. <u>DIFFUSE RADIATION</u> Clouds and particles i the atmosphere not only absorb solar energy but scatter it in all directions. As a result, part of the solar radiation reaching the earth' surface is diffused, and received from all par of the sky. Diffuse radiation, as opposed t direct radiation, is more predominant on haz days than clear ones. At most, however, diffus radiation can only be about one quarter of th solar constant, or about 100 BTU/hr./sq. ft.

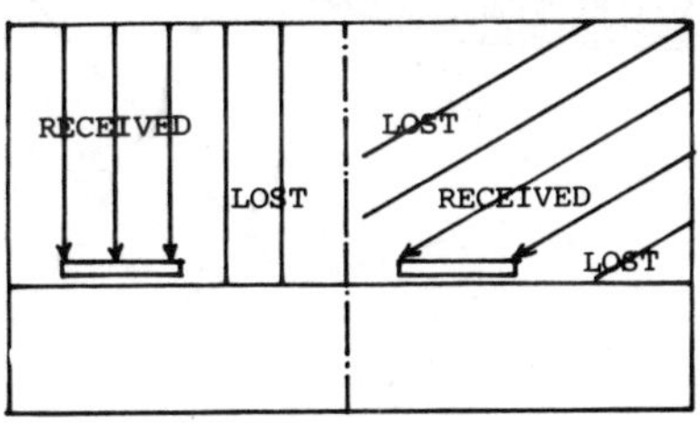

5. <u>DIRECT RADIATION ON A HORIZONT SURFACE</u> Although the amount of radiati remains constant, less radiation strikes a giv horizontal area as the sun gets lower in the sk

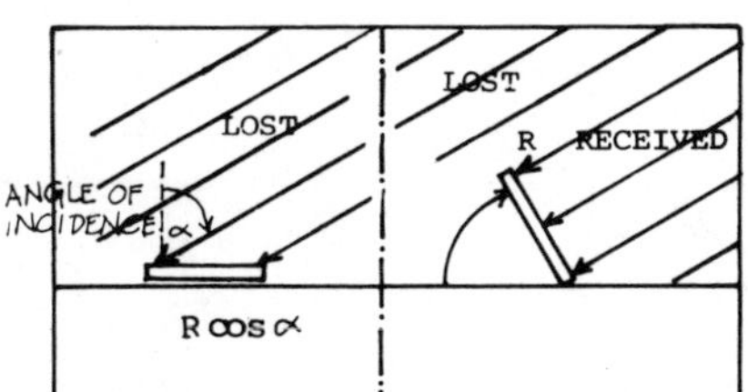

6. <u>DIRECT SOLAR RADIATION ON A TILTE SURFACE</u> The same principle applies to tilted surface such as a collector. By tilting th collector so that it is nearly perpendicular t the sun's ray, more energy strikes its surfac undiminished by a cosine factor.

C. Solar Window

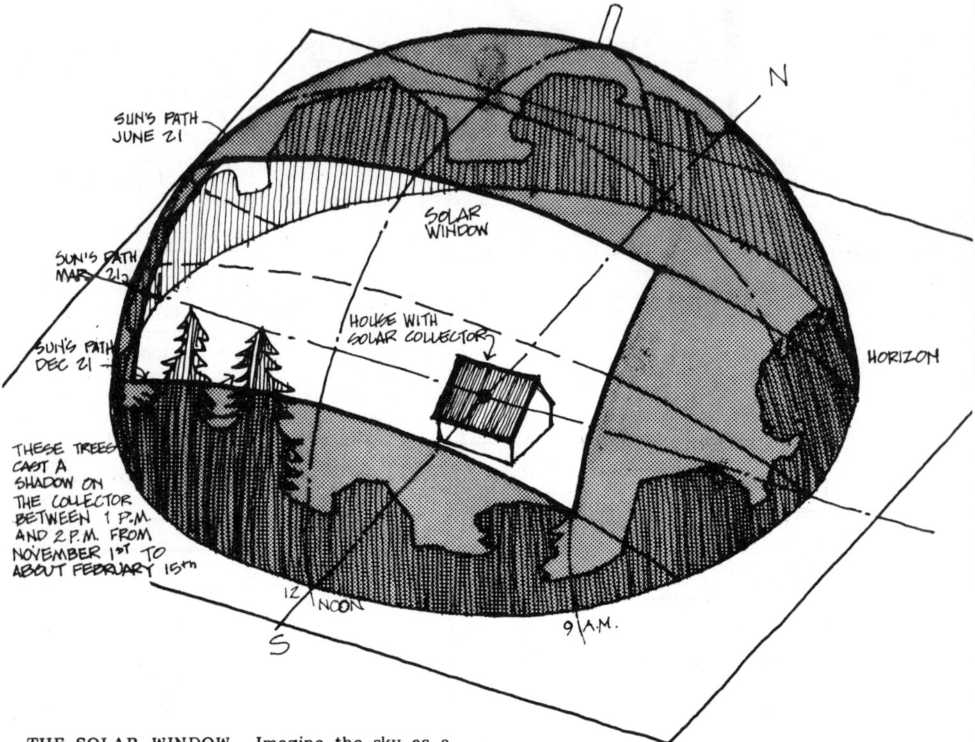

<u>THE SOLAR WINDOW</u> Imagine the sky as a transparent dome with its center at the solar collector of a house. The path of the sun can be painted (projected) onto the dome, as can be the outline of surrounding houses and trees The morning and afternoon limits of useful solar collection (roughly 9 A.M. and 3 P.M.) and the sun's path between those hours throughout the year scribes a "solar window" on the dome. Almost all of the useful sun that reaches the collector must come through this window except for the added effect of diffuse radiation. If any of the surrounding houses, trees, etc., intrude into this "solar window," the intrusion will cast a shadow on the collector. The isometric drawing above illustrates the "solar window" for a latitude 40° N. The solar window will change for different latitudes.

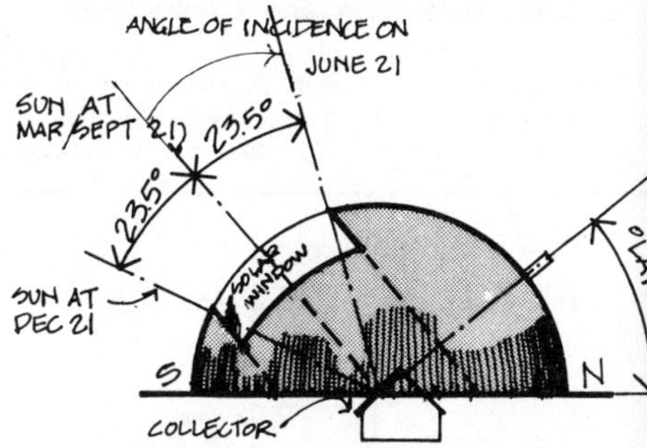

SIDE VIEW OF SKY DOME WITH "SOLAR WINDOW" A side view of the sky dome from the east illustrates the relative position and angle of the sun throughout the year that defines the boundaries of the "solar window."

ANGLE OF INCIDENCE, a term often used in solar collector design, is the angle measured from t normal of the collector surface to the line indicating the sun's altitude at a particular time. T diagram specifically identifies the angle of incidence for June 21.

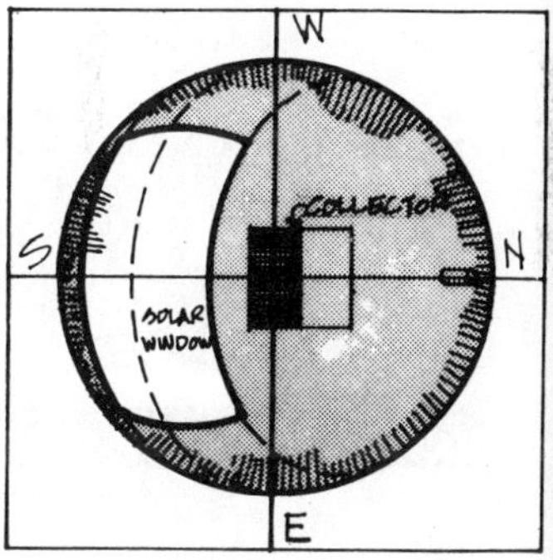

PLAN VIEW OF SKY DOME WITH "SOLAR WINDOW" Viewed from above the sky dome, the seasonal path of the sun can be plotted thus defining the boundaries of the "solar window." This is easily accomplished by the use of a standard sun path diagram for the proper latitude. Sun path diagrams are widely reproduced and used for determining the azimuth and altitude of the sun at any time during the year, and give the points which can be plotted to determine the solar window.

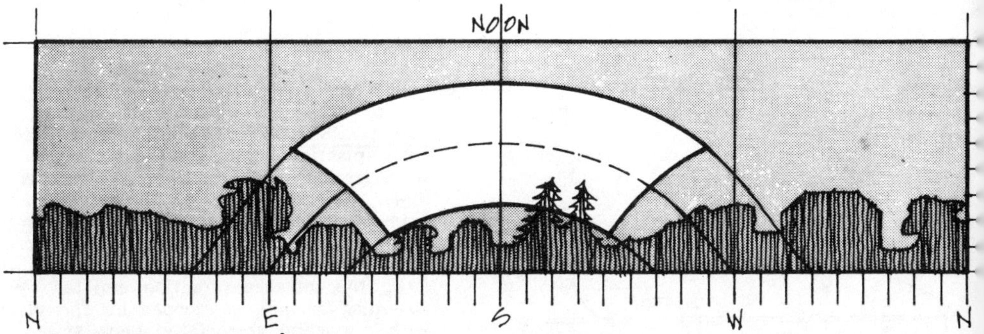

PANORAMA OF THE SKY DOME As with the spherical earth, the spherical sky dome with i "solar window" can be mapped using a Mercator projection, in which all latitude and longitude lin are straight lines. Such a map is very useful for comparing the site surroundings with the "sol window" outline, since both can be easily plotted on the map. Any elements surrounding the si that intrude into the "solar window" will cast shadows on the collector.

D. Solar Collection and Conversion

Basic to the utilization of solar energy for space and domestic hot water heating is the process by which solar radiation is converted to thermal energy. This conversion process is the basic link between the energy supply -- the sun — and the energy load -- the dwelling. The process is best understood by briefly explaining solar radiation and then discussing the characteristics of collection.

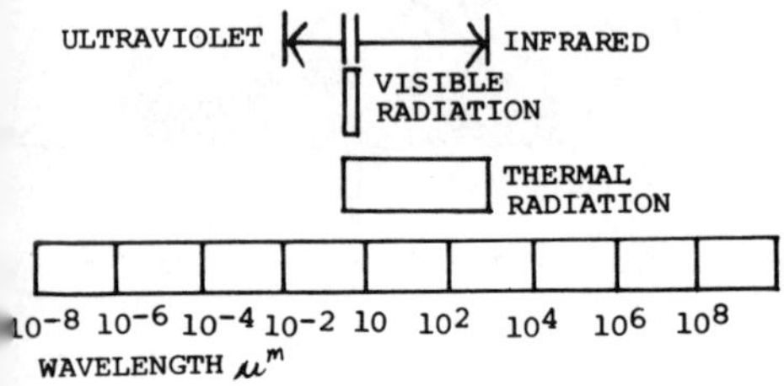

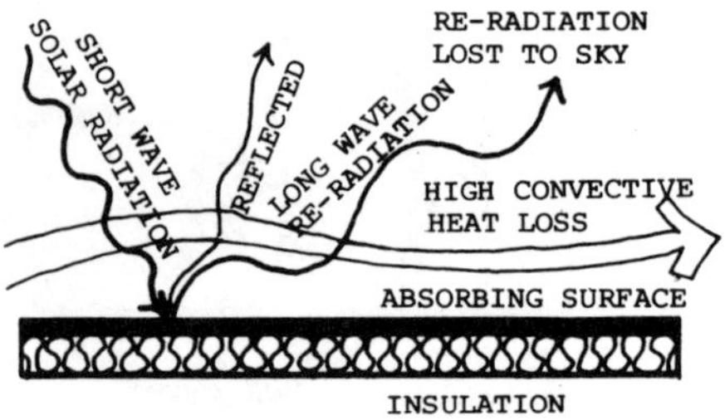

1. SOLAR ENERGY CONVERSION Solar radiation is electromagnetic radiation generated by the sun, which reaches the earth's surface with a wavelength distribution of .3 to 2.4 micrometers. Radiation is perceived as visible light between ultraviolet and infrared, specifically between .36 and .76 micrometers. For most solar applications, solar radiation in the visible and near-infrared range is the most important.

The drawings to the left show the principle of solar energy collection and conversion. When incoming solar radiation impinges on the surface of a body, it is partially absorbed, partially reflected, and, if the body is transparent, partially transmitted. The relative magnitude of each varies with the surface characteristics, body geometry, material composition, and wavelength.

For solar applications, energy must be first absorbed, then converted into thermal energy and, finally removed by a heat transfer mechanism in order to be useful. Absorbed radiation heats up the absorbing body, which then re-emits energy in the form of thermal radiation in the infrared (longwave) part of the spectrum. If the absorbing surface is exposed to the atmosphere, part of the absorbed energy will be lost by convection or radiation.

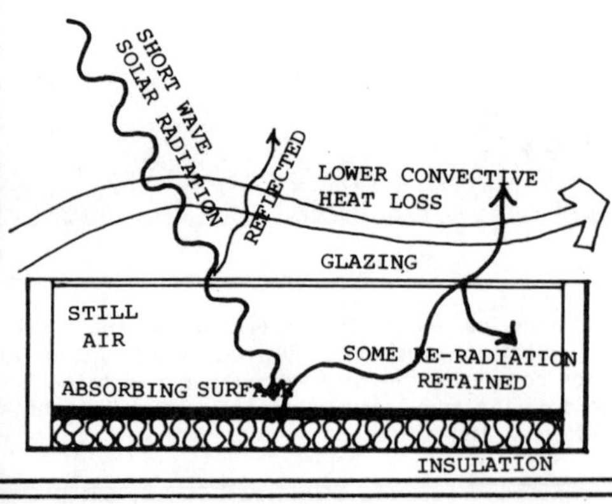

2. THE GREENHOUSE EFFECT Most glass and some plastics are transparent in the solar wavelength region and hence are used as windows. At the same time, this glazing has low transmission in the infrared (longwave) region. By placing glass or plastic over the absorber in a collector, energy is trapped in two ways: first, the infrared radiation emitted by the absorbing surface is stopped by the glazing, with a portion re-radiated back toward the absorber, and thereby trapped. Second, the glazing also traps a layer of still air next to the absorber and reduces the convective heat loss. This behavior of glazing is called the "greenhouse effect" and is used in most solar collectors.

F. Shading of Collector

Another issue related to both collector orientation and tilt is shading. Solar collectors should be located on the building or site so that unwanted shading of the collectors by adjacent structures, landscaping or building elements does not occur. In addition, considerations for avoiding shading of the collector by other collectors should also be made. Collector shading by elements surrounding the site may be addressed by considering the "solar window" concept.

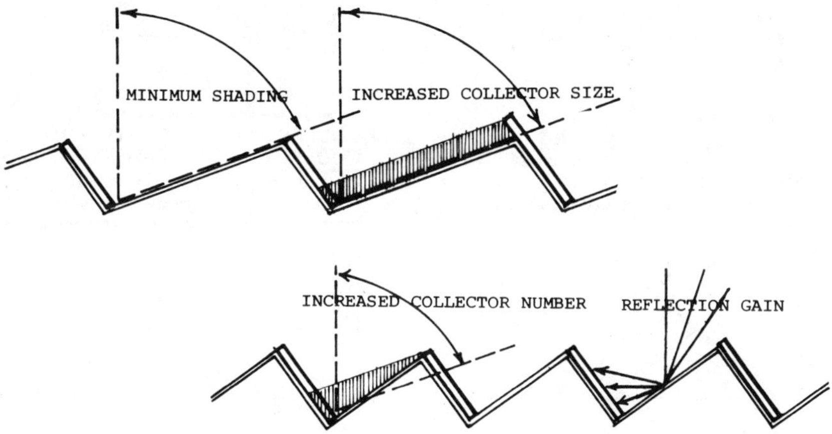

1. SELF-SHADING OF COLLECTOR Avoiding all shelf-shading for a bank of parallel collectors during useful collection hours (9 AM and 3 PM) results in designing for the lowest angle of incidence with large spaces between collectors. It may be desirable therefore to allow some self-shading at the end of solar collection hours, in order to increase collector size or to design a closer spacing of collectors, thus increasing solar collection area. By making the collector's back slope reflective, one could increase the amount of solar radiation striking the adjacent collector, thus negating some of the shading loss.

SHADING OF COLLECTOR BY BUILDING ELEMENTS Chimneys, parapets, fire walls, dormers, and other building elements can cast shadows on adjacent roof-mounted solar collectors, as well as on vertical wall collectors. The drawing to the right shows a house with a 45° south-facing collector at latitude 40° North. By mid-afternoon portions of the collector are shaded by the chimney, dormer, and the offset between the collector on the garage. Careful attention to the placement of building elements and to floor plan arrangement is required to assure that unwanted collector shading does not occur.

E. Collector Orientation and Tilt

Solar collectors must be oriented and tilted within prescribed limits to receive the optimum level c solar radiation for system operation and performance.

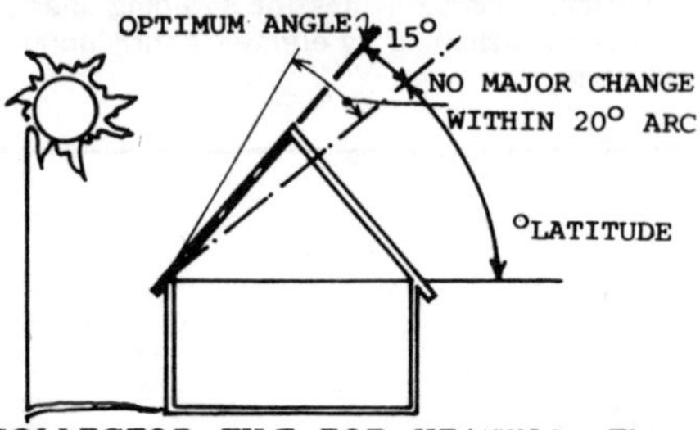

1. COLLECTOR TILT FOR HEATING The optimum collector tilt for heating is usually equal to the site latitude plus 10 to 15 degrees. Variations of 10 degrees on either side of this optimum are acceptable.

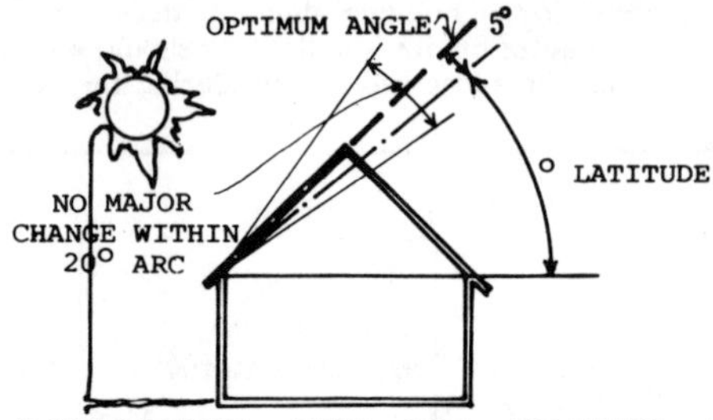

2. COLLECTOR TILT FOR HEATING AN COOLING The optimum collector tilt f heating and cooling is usually equal to si latitude plus 5 degrees. Variations of 10 degre on either side of the optimum are acceptable.

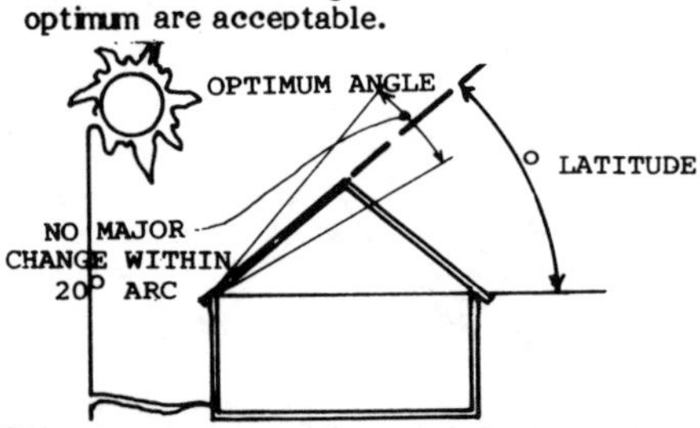

3. COLLECTOR TILT FOR DOMESTIC HOT WATER The optimum collector tilt for domestic water heating alone is usually equal to the site latitude. Again, variations of 10 degrees on either side of the optimum are acceptable.

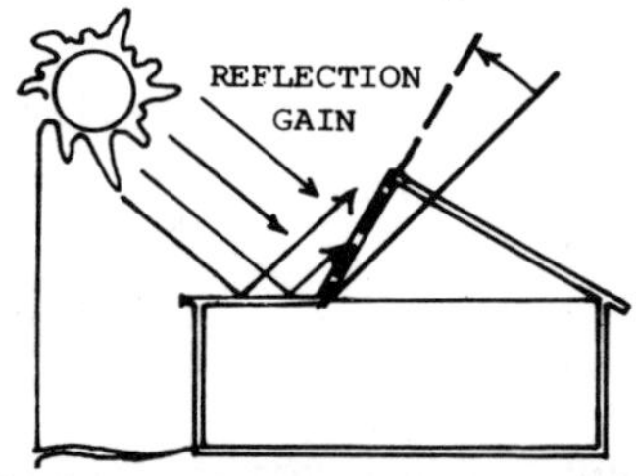

4 MODIFICATION OF OPTIMUM COLLECTO TILT A greater gain in solar radiation col lection sometimes may be achieved by tiltir the collector away from the optimum in order t capture radiation reflected from adjacer ground or building surfaces. The correspondin reduction of radiation directly striking th collector, due to non-optimum tilt, should b recognized when considering this option.

5. SNOWFALL CONSIDERATION The snowfall characteristics of an area may influence th appropriateness of these optimum collector tilts. Snow buildup on the collector, or drifting in fror of the collector, should be avoided.

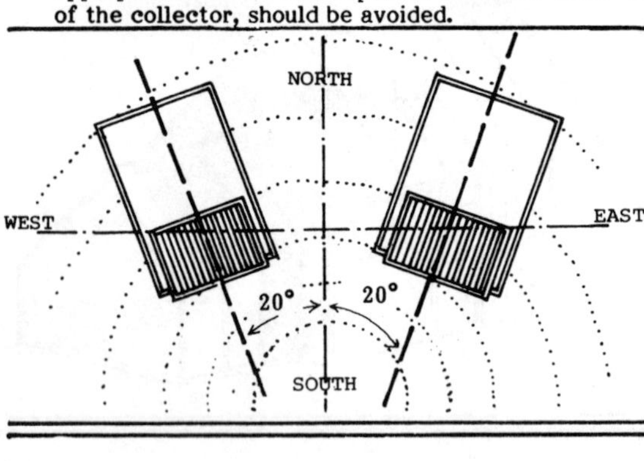

COLLECTOR ORIENTATION A collector c ientation of 20 degrees to either side of tr South is acceptable. However, local clima and collector type may influence the choi between East or West deviations.

Selected References

Becker, Ralph E., *Common Law Sun Rights: An Obstacle to Solar Heating and Cooling,* JOURNAL OF CONTEMPORARY LAW 3 (1976): 19–36.

Bersohn, David L., *Securing Solar Energy Rights: Easement, Nuisance, or Zoning?,* COLUMBIA JOURNAL OF ENVIRONMENTAL LAW 3 (1976): 112–52.

Bezdek, Roger, et al., ANALYSIS OF POLICY OPTIONS FOR ACCELERATING COMMERCIALIZATION OF SOLAR HEATING AND COOLING SYSTEMS. Washington, D.C.: George Washington University, Behavioral Studies Group, 1977.

Dubin, Fred S., ANALYSIS OF ENERGY USAGE ON LONG ISLAND FROM 1975 TO 1995: THE OPPORTUNITIES TO REDUCE PEAK ELECTRICAL DEMANDS AND ENERGY CONSUMPTION BY ENERGY CONSERVATION, SOLAR ENERGY, WIND ENERGY, AND TOTAL ENERGY SYSTEMS. Suffolk County, N.Y.: Suffolk County Department of Environmental Control, 1975.

Eisenstadt, Melvin M.; Long, Stephen C. M.; and Utton, Albert E., *A Proposed Solar Zoning Ordinance,* URBAN LAW ANNUAL 15 (1978): 211–250.

Eisenstadt, Melvin M.; Long, Stephen C. M.; and Utton, Albert E., *A Proposed Solar Zoning Ordinance.* Mimeographed. Prepared for the New Mexico Energy Institute. ERB-76-103, November 1977.

Eisenstadt, Melvin M.; Long, Stephen C. M.; and Utton, Albert E., *Solar Rights and Their Effect on Solar Heating and Cooling,* NATURAL RESOURCES JOURNAL 16 (1976): 363–414.

Environmental Law Institute, LEGAL BARRIERS TO SOLAR HEATING AND COOLING OF BUILDINGS. Prepared for the U.S. Department of Energy. Springfield, Va.: National Technical Information Service, HCP/M 2528-01, March 1978.

Gevurtz, Franklin, *Obstruction of Sunlight as a Private Nuisance,* CALIFORNIA LAW REVIEW 65 (1977): 94–119.

Harris, William R., *Is the Right to Light a California Necessity?* Statement before the California State Assembly Committee on the Judiciary. Santa Monica, Cal.: Rand Corporation, 1975.

Hillhouse, Karin H., LEGAL AND INSTITUTIONAL PERSPECTIVES ON SOLAR ENERGY IN COLORADO: A CASE STUDY OF LAND USE AND ENERGY DECISION-MAKING. Washington, D.C.: Environmental Law Institute, 1977.

Jaffe, Martin, *Some Comments on Drafting Solar Access Regulations,* LAND USE LAW AND ZONING DIGEST 30 (1978): 4–7.

Johnson, Milan, and Tiedemann, Thomas F., SOLAR ENERGY COMMERCIALIZATION AT THE STATE LEVEL: THE FLORIDA SOLAR ENERGY WATER HEATER PROGRAM. Report to the Federal Energy Administration. Cape Canaveral: Florida Solar Energy Center, FSEC-76-3, March 1977.

Knowles, Ralph, *Solar Ethics—Urban Form.* Paper delivered at the Passive Solar Conference, Philadelphia, Pa., 16 March 1978.

Kraemer, Sandy F., SOLAR LAW. Colorado Springs: Shepard's, Inc., 1978.

Miller, Alan S., *Solar Energy and Land Use in Colorado.* ENVIRONMENTAL LAW REPORTER 6 (1976): 50039–46.

Miller, Stephen S., *Let the Sunshine In: A Comparison of Japanese and American Solar Rights,* HARVARD ENVIRONMENTAL LAW REVIEW 1 (1976): 578–600.

Moskowitz, Daniel P., *Legal Access to Light: The Solar Energy Imperative,* NATURAL RESOURCES LAWYER 9 (1976): 177–208.

Olgyay, Victor, DESIGN WITH CLIMATE: BIOCLIMATIC APPROACH TO ARCHITECTURAL REALISM. Princeton, N.J.: Princeton University Press, 1963.

Ontario Ministry of Energy, PERSPECTIVES ON ACCESS TO SUNLIGHT. Toronto, Ontario, May 1978.

Phillips, James D., ASSESSMENT OF A SINGLE-FAMILY RESIDENCE SOLAR HEATING SYSTEM IN A SUBURBAN DEVELOPMENT SETTING. Prepared with the support of the National Science Foundation. Springfield, Va: National Technical Information Service, NSF/RA-760192, July 1976.

Reitze, Arnold W., Jr., and Reitze, Glenn L., *Protecting a Place in the Sun (Part One).* ENVIRONMENT, June 1976, pp. 2–4.

Reitze, Arnold W., Jr., and Reitze, Glenn L , *Protecting a Place in the Sun (Part Two).* ENVIRONMENT, July–August 1976, p. 4.

Reitze, Glenn L., *Solar Rights Zoning Guarantee: Seeking New Law In Old Concepts,* WASHINGTON UNIVERSITY LAW QUARTERLY 1976: 378–402.

Rivkin, Steven K., *Sun Rights and Land Use,* in EARLY USE OF SOLAR ENERGY IN BUILDINGS. Prepared for the National Science Foundation. Washington, D.C.: American Institute of Architects Research Corporation, 1976.

Robbins, Richard L., *Law and Solar Energy Systems: Legal Impediments and Inducements to Solar Energy Systems,* SOLAR ENERGY 18 (1976): 371–79.

SOLAR LAW REPORTER. Golden, Colo.: Solar Energy Research Institute, to commence publication in spring 1979.

Takagi, Gail Feingold, *Designs on Sunshine: Solar Access in the United States and Japan,* CONNECTICUT LAW REVIEW 10 (1978): 123–46.

Thomas, William A.; Miller, Alan S.; and Robbins, Richard L., OVERCOMING LEGAL UNCERTAINTIES ABOUT USE OF SOLAR ENERGY SYSTEMS. Chicago: American Bar Foundation, 1978.

U.S. Congress, Office of Technology Assessment, APPLICATION OF SOLAR TECHNOLOGY TO TODAY'S ENERGY NEEDS. Vol. 1. Washington, D.C.: U.S. Government Printing Office, June 1978.

Urban Land Institute, Solar Energy and Land Use issue of ENVIRONMENTAL COMMENT, May 1978.

White, Mary D., *The Allocation of Sunlight: Solar Rights and the Prior Appropriation Doctrine,* UNIVERSITY OF COLORADO LAW REVIEW 47 (1976): 421–47.

Wilson, Jones, Morton & Lynch, Santa Clara, California, COMMUNITY CENTER COMMERCIAL SOLAR DEMONSTRATION: LEGAL ALTERNATIVES, IMPLICATION, AND FINANCING OF SOLAR HEATING AND COOLING BY A MUNICIPAL CORPORATION. Santa Clara, Cal.: City of Santa Clara, 1976.

Zillman, Donald N., and Deeny, Raymond, *Legal Aspects of Solar Energy Development,* ARIZONA STATE LAW JOURNAL 1976: 25–58.

Index

About the Author

Gail Boyer Hayes, a lawyer, is a senior associate with the Environmental Law Institute. She has written and spoken extensively in the field of solar law, and co-authored an earlier ELI book, *Legal Barriers to Solar Heating and Cooling of Buildings.* A former journalist, she has written articles for national periodicals and co-hosted the "City Watchers" television show in Los Angeles.